The CD accompanying the print version is no longer available.

Please see Appendix A, page 273
for instructions on where to download free versions
of the software discussed in this book.

TRANSMISSION LINES IN DIGITAL AND ANALOG ELECTRONIC SYSTEMS

TRANSMISSION LINES IN DIGITAL AND ANALOG ELECTRONIC SYSTEMS

Signal Integrity and Crosstalk

CLAYTON R. PAUL
Department of Electrical and Computer Engineering
Mercer University
Macon, Georgia
and
Emeritus Professor of Electrical Engineering
University of Kentucky
Lexington, Kentucky

IEEE PRESS

WILEY

A JOHN WILEY & SONS, INC., PUBLICATION

Library of Congress Cataloging-in-Publication Data:

Paul, Clayton R.
  Transmission lines in digital and analog electronic systems : signal integrity and
crosstalk / Clayton R. Paul.
     p. cm.
    ISBN 978-0-470-59230-4
  1. Multiconductor transmission lines.   2. Telecommunication lines.   3. Crosstalk.
4. Signal integrity (Electronics)   5. Electronic circuits.   I. Title.
   TK872.T74P3824 2010
   621.382'3–dc22

                                                                2010006495

10 9 8 7 6 5 4 3 2 1

*This book is dedicated to the humane and compassionate
treatment of animals*

and my beloved pets:

*Patsy, Dusty, Megan, Tinker, Bunny, Winston, Sweetheart, Lady, Tigger,
Beaver, Ditso, Buru, Old Dog, Zip, Tara, Timothy, Kiko, Valerie, Red,
Sunny, Johnny, Millie, Molly, Angel, Autumn, and Shabby.*

*Those readers who are interested in the humane and compassionate
treatment of animals are encouraged to donate to*

*The Clayton and Carol Paul Fund for Animal Welfare
c/o the Community Foundation of Central Georgia
277 MLK, Jr. Blvd
Suite 303
Macon, GA 31202*

The primary and only objective of this Fund is to provide monetary grants to

(1) animal humane societies
(2) animal shelters
(3) animal adoption agencies
(4) low-cost spay-neuter clinics
(5) individual wildlife rehabilitators
(6) as well as other organizations devoted to animal welfare

in order to allow these volunteer organizations to use their enormous enthusiasm,
drive and willingness to reduce animal suffering and homelessness through the
monetary maintenance of their organizations where little or no monetary funds
existed previously.

CONTENTS

Preface   **xi**

**1 Basic Skills and Concepts Having Application
   to Transmission Lines**   **1**

   1.1   Units and Unit Conversion   3
   1.2   Waves, Time Delay, Phase Shift, Wavelength, and
         Electrical Dimensions   6
   1.3   The Time Domain vs. the Frequency Domain   11
         1.3.1   Spectra of Digital Signals   12
         1.3.2   Bandwidth of Digital Signals   17
         1.3.3   Computing the Time-Domain Response
                 of Transmission Lines Having Linear
                 Terminations Using Fourier Methods and
                 Superposition   27
   1.4   The Basic Transmission-Line Problem   31
         1.4.1   Two-Conductor Transmission Lines and Signal
                 Integrity   32
         1.4.2   Multiconductor Transmission Lines and
                 Crosstalk   41
   Problems   46

PART I   TWO-CONDUCTOR LINES AND SIGNAL INTEGRITY   49

2   Time-Domain Analysis of Two-Conductor Lines   51

2.1   The Transverse Electromagnetic (TEM) Mode
of Propagation and the Transmission-Line Equations   52
2.2   The Per-Unit-Length Parameters   56
   2.2.1   Wire-Type Lines   57
   2.2.2   Lines of Rectangular Cross Section   68
2.3   The General Solutions for the Line Voltage and Current   71
2.4   Wave Tracing and Reflection Coefficients   74
2.5   The SPICE (PSPICE) Exact Transmission-Line Model   84
2.6   Lumped-Circuit Approximate Models of the Line   91
2.7   Effects of Reactive Terminations on Terminal
Waveforms   92
   2.7.1   Effect of Capacitive Terminations   92
   2.7.2   Effect of Inductive Terminations   94
2.8   Matching Schemes for Signal Integrity   96
2.9   Bandwidth and Signal Integrity: When Does
the Line Not Matter?   104
2.10   Effect of Line Discontinuities   105
2.11   Driving Multiple Lines   111
Problems   113

3   Frequency-Domain Analysis of Two-Conductor Lines   121

3.1   The Transmission-Line Equations for Sinusoidal
Steady-State Excitation of the Line   122
3.2   The General Solution for the Terminal Voltages
and Currents   123
3.3   The Voltage Reflection Coefficient and Input Impedance
to the Line   123
3.4   The Solution for the Terminal Voltages and Currents   125
3.5   The SPICE Solution   128
3.6   Voltage and Current as a Function of Position
on the Line   130
3.7   Matching and VSWR   133
3.8   Power Flow on the Line   134
3.9   Alternative Forms of the Results   137
3.10   The Smith Chart   138
3.11   Effects of Line Losses   147

3.12 Lumped-Circuit Approximations for Electrically
      Short Lines   161
3.13 Construction of Microwave Circuit Components Using
      Transmission Lines   167
Problems   170

PART II THREE-CONDUCTOR LINES AND CROSSTALK   175

4 The Transmission-Line Equations for Three-Conductor Lines 177

4.1 The Transmission-Line Equations for Three-Conductor Lines 177
4.2 The Per-Unit-Length Parameters   184
     4.2.1 Wide-Separation Approximations for Wires   185
     4.2.2 Numerical Methods   196
Problems   205

**5 Solution of the Transmission-Line Equations
    for Three-Conductor Lossless Lines   207**
5.1 Decoupling the Transmission-Line Equations with Mode
     Transformations   208
5.2 The SPICE Subcircuit Model   210
5.3 Lumped-Circuit Approximate Models of the Line   227
5.4 The Inductive-Capacitive Coupling Approximate Model   232
Problems   236

**6 Solution of the Transmission-Line Equations
    for Three-Conductor Lossy Lines   239**
6.1 The Transmission-Line Equations for Three-Conductor
     Lossy Lines   240
6.2 Characterization of Conductor and Dielectric Losses   244
     6.2.1 Conductor Losses and Skin Effect   244
     6.2.2 Dielectric Losses   248
6.3 Solution of the Phasor (Frequency-Domain)
     Transmission-Line Equations for a
     Three-Conductor Lossy Line   251
6.4 Common-Impedance Coupling   260
6.5 The Time-Domain to Frequency-Domain Method   261
Problems   270

Appendix   A Brief Tutorial on Using PSPICE   273

Index   295

PREFACE

This book is intended as a textbook for a senior/first-year graduate-level course in transmission lines in electrical engineering (EE) and computer engineering (CpE) curricula. It has been class tested at the author's institution, Mercer University, and contains virtually all the material needed for a student to become competent in all aspects of transmission lines in today's high-frequency analog and high-speed digital world. The book is also essential for industry professionals as a compact review of transmission-line fundamentals.

Until as recently as a decade ago, digital system clock speeds and data rates were in the hundreds of megahertz range. Prior to that time, the "lands" on printed circuit boards (PCBs) that interconnect the electronic modules had little or no impact on the proper functioning of those electronic circuits. Today, the clock and data speeds have moved into the low gigahertz range. As the demand for faster data processing continues to escalate, these speeds will no doubt continue to increase into the gigahertz frequency range. In addition, analog communication frequencies have also moved steadily into the giga-hertz range and will no doubt continue to increase. Although the physical dimensions of these lands and the PCBs supporting them have not changed significantly over these intervening years, the spectral content of the signals they carry has increased significantly. Because of this the electrical dimensions (in wavelengths) of the lands have increased to the point where these interconnects have a significant effect on the signals they are carrying, so that just getting the systems to work properly has become a major design problem. This has generated a new design problem, referred to as *signal integrity*. Good signal integrity means that the interconnect conductors should not adversely affect the operation of the modules that the conductors interconnect. Prior to some 10 years ago, these interconnects could be modeled reliably with lumped-circuit models that are easily analyzed using Kirchhoff's voltage and current laws and other lumped-circuit analysis methods. Because these interconnects are becoming "electrically long," lumped-circuit modeling of them is becoming inadequate and gives erroneous answers. Most interconnect conductors must now be treated as distributed-circuit *transmission lines.*

In the last 30 years there have been dramatic changes in electrical technology, yet the length of the undergraduate curriculum has remained four years. Since the undergraduate curriculum is a "zero-sum game", the introduction of courses necessitated by the advancements in technology, in particular digital technology, has caused many of the standard topics to disappear from the curriculum or be moved to senior technical electives which not all graduates take. The subject of transmission lines is an important example of this. Until a decade ago, the analysis of transmission lines was a standard topic in the EE and CpE undergraduate curricula. Today most of the undergraduate curricula contain a rather brief study of the analysis of transmission lines in a one-semester junior-level course on electromagnetics (often the only course on electromagnetics in the required curriculum). In some schools, this study of transmission lines is relegated to a senior technical elective or has disappeared from the curriculum altogether. This raises a serious problem in the preparation of EE and CpE undergraduates to be competent in the modern industrial world. For the reasons mentioned above, today's undergraduates lack the basic skills to design high-speed digital and high-frequency analog systems. It does little good to write sophisticated software if the hardware is unable to process the instructions. This problem will increase as the speeds and frequencies of these systems continue to increase, seemingly without bound. This book is meant to repair that basic deficiency.

In Chapter 1, the fundamental concepts of waves, wavelength, time delay, and electrical dimensions are discussed. In addition, the bandwidth of digital signals and its relation to pulse rise and fall times is discussed. Preliminary discussions of *signal integrity* and *crosstalk* are also given.

Part I contains two chapters covering two-conductor transmission lines and designing for *signal integrity*. Chapter 2 covers the time-domain analysis of those transmission lines. The transmission-line equations are derived and solved, and the important concept of characteristic impedance is covered. The important per-unit-length parameters of inductance and capacitance that distinguish one line from another are obtained for typical lines. The terminal voltages and currents of lines with various source waveforms and resistive terminations are computed by hand via wave tracing. This gives considerable insight into the general behavior of transmission lines in terms of forward- and backward-traveling waves and their reflections. The SPICE computer program and its personal computer version, PSPICE, contain an *exact* model for a two-conductor lossless line and is discussed as a computational aid in solving for transmission-line terminal voltages and currents. SPICE is an important computational tool since it provides a determination of the terminal voltages and currents for practical linear and nonlinear terminations such as CMOS and bipolar devices, for which hand analysis is very formidable. Matching schemes for achieving signal integrity are covered, as are the effects of line discontinuities. Chapter 3 covers the corresponding analysis in the frequency domain. The important analog concepts of input impedance to the line, VSWR and the Smith chart (which provides considerable insight), are also discussed. The effect of line losses, including skin effect in the line conductors and dielectric losses in the surrounding dielectric, are becoming increasingly critical, and their detrimental effects are discussed.

Part II repeats these topics for three-conductor lines in terms of the important detrimental effects of *crosstalk* between transmission lines. Crosstalk is becoming of paramount concern in the design of today's high-speed and high-frequency electronic systems. The transmission-line equations for three-conductor *lossless* lines are derived, and the important per-unit-length matrices of the inductance and capacitance of the lines are covered in Chapter 4. Numerical methods for computing the per-unit-length parameter matrices of inductance and capacitance are studied, and computer programs are given that compute these numerically for ribbon cables and various structures commonly found on PCBs. Chapter 5 covers the solution of three-conductor *lossless* lines via mode decoupling. A SPICE subcircuit model is determined via this decoupling and implemented in the computer program SPICEMTL.EXE. This program performs the tedious diagonalization of the per-unit-length parameter matrices and gives as output a SPICE subcircuit for modeling lossless coupled lines. As in the case of two-conductor lines, this allows the study of line responses not only for resistive loads but, more important, nonlinear and/or reactive loads such as CMOS and bipolar devices that are common line terminations in today's digital systems. How to incorporate the frequency-dependent losses of the line conductors and the surrounding dielectric into a solution for the crosstalk voltages is discussed in Chapter 6. The frequency-domain solution of the MTL equations is again given in terms of similarity transformations in the frequency domain. The time-domain solution for the crosstalk voltages is obtained in terms of the frequency-domain transfer function, which is obtained by superimposing the responses to the Fourier components of $V_S(t)$.

The appendix gives a brief tutorial of SPICE (PSPICE), which is used extensively throughout the book. Several computer programs used and described in this book for computing the per-unit-length parameter matrices and a subcircuit model for three-conductor lines are contained in a CD that is included with the book along with two MATLAB programs for computing the Fourier components of a digital waveform. The CD also contains two versions of PSPICE.

Each chapter concludes with numerous problems for the reader to practice his or her understanding of the material. The answers to those that are simply stated are given in brackets, [·], at the end of the question. The answers to most of the other problems can be verified using PSPICE. In those cases, the hand calculations should be checked using PSPICE. If these disagree, there is an error in either (1) the hand calculation, (2) the PSPICE setup, or (3) both. In this case, the reader should determine the error so that both answers agree. Getting the hand calculations and those obtained with PSPICE to agree is a tremendously useful learning tool.

This book grew out of the realization that most of today's EE and CpE graduates lack a critically important skill: the analysis of transmission lines. If we, as educators, are to prepare our graduates adequately for the increasingly difficult design problems of a high-speed digital world, it is imperative that we institute a dedicated course devoted to the analysis of transmission lines. This book is devoted to achieving that objective.

CLAYTON R. PAUL

Macon, Georgia

1

BASIC SKILLS AND CONCEPTS HAVING APPLICATION TO TRANSMISSION LINES

We are rapidly moving into a *digital era*. Until as recently as some 10 years ago, clock speeds and data rates of digital systems were in the hundreds of megahertz (MHz) range, with rise and fall times of the pulses in the nanosecond (1 ns $= 10^{-9}$ s) range. Prior to that time, the *lands* (conductors of rectangular cross section) that interconnect the electronic modules on printed circuit boards (PCBs) had little or no impact on the proper functioning of those electronic circuits. Today, the clock and data speeds have moved rapidly into the low-gigahertz (GHz) range. The rise and fall times of those digital waveforms have decreased into the picosecond (1 ps $= 10^{-12}$ s) range. For example, a 1-GHz digital clock signal consists of trapezoidal-shaped pulses having rise and fall times on the order of 100 ps or less. A digital clock waveform is illustrated in Fig. 1.1.

The period of the periodic waveform T is the reciprocal of the clock fundamental frequency f_0. The rise and fall times are denoted as τ_r and τ_f, respectively, and the pulse width (between 50% levels) is denoted as τ. Digital data waveforms are similar except that the period starts immediately after the previous pulse [i.e. $T = \tau + (\tau_r + \tau_f)/2$], and the occurrence of a pulse during the adjacent time intervals is random. As the frequencies of the clocks increase, the period T decreases and hence the rise and fall times of the pulses must be reduced commensurately in order that pulses resemble a trapezoidal

Transmission Lines in Digital and Analog Electronic Systems: Signal Integrity and Crosstalk, By Clayton R. Paul
Copyright © 2010 John Wiley & Sons, Inc.

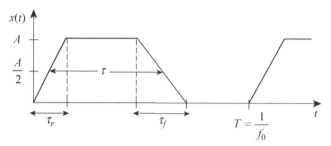

FIGURE 1.1. Typical digital clock/data waveform.

shape rather than a sawtooth waveform, thereby giving adequate setup and hold time intervals. Reducing the pulse rise and fall times has had the consequence of increasing the spectral content of the waveshape according to the Fourier series of the waveform. Typically, this spectral content is significant up to the inverse of the rise and fall times, $1/\tau_r$, as we will see.

For example, a 1-GHz digital clock signal having rise and fall times of 100 ps has significant spectral content at multiples (harmonics) of the basic clock frequency (1 GHz, 2 GHz, 3 GHz, ...) up to around 10 GHz. As the demand for faster data processing continues to escalate, these speeds will no doubt continue to increase into the gigahertz frequency range. The pulse rise and fall times will be reduced commensurately, thereby increasing the spectral content further into the gigahertz frequency range. This also applies to mixed-signal systems containing both digital and analog signals.

Although the *physical lengths* of the lands that interconnect the electronic modules on the PCBs have not changed significantly over these intervening years, their *electrical lengths* (in wavelengths) have increased because of the increased spectral content of the signals that the lands carry. Today these *interconnects* can have a significant effect on the signals they are carrying, so that just getting the systems to work properly has become a major design problem. Remember that it does no good to write sophisticated software if the hardware cannot execute those instructions faithfully. This has generated a new design problem, referred to as *signal integrity*. Good signal integrity means that the interconnect conductors (the lands) should not adversely affect the operation of the modules that the conductors interconnect.

Prior to some 10 years ago, these interconnects could be modeled reliably with lumped-circuit models that are easily analyzed using Kirchhoff's voltage and current laws and lumped-circuit analysis methods, or could be ignored altogether. Because these interconnects are becoming "electrically long," lumped-circuit modeling of them is becoming inadequate and gives erroneous answers. The interconnect conductors must now be treated as distributed-circuit *transmission lines*. The interaction of the electric and magnetic fields between two adjacent transmission lines also causes portions of the voltage and current waveforms on one line to appear inadvertently at the ends of the adjacent line, thereby creating potential interference problems in the electronic devices that the adjacent line interconnects. This is called *crosstalk* and is also rapidly becoming a significant problem in high-speed digital electronics.

This book is intended to be a thorough but concise description of the analysis of transmission lines with respect to signal integrity and crosstalk in modern high-speed digital and high-frequency analog systems. This chapter covers some important basic skills and concepts that facilitate an understanding of the behavior of those transmission lines as well as demonstrating when the interconnects need to be considered as transmission lines.

1.1   UNITS AND UNIT CONVERSION

The internationally accepted system of units is the International System, or *SI system*, where the primary units are the meter, kilogram, second, and ampere, thus termed the *MKSA system*. *All quantities in any formula or law must be in these units.* For example, Coulomb's law for the force between two point charges that are separated a distance R is

$$F = \frac{Q_1 Q_2}{4\pi \varepsilon_0 R^2} \qquad \text{N(newtons)}$$

The SI units of the charges Q_1 and Q_2 are coulombs (C), and the distance between the two charges, R, is in meters (m). The SI units of force are newtons (N). The constant in the denominator is the *permittivity* of free space (essentially air):

$$\varepsilon_0 \cong \frac{1}{36\pi} \times 10^{-9} \text{F/m}$$

We will see in the other electromagnetic laws a similar constant, known as the *permeability* of free space:

$$\mu_0 = 4\pi \times 10^{-7} \text{H/m}$$

These two constants have the units of a capacitance in farads (F) per unit of length and an inductance in henrys (H) per unit of length. These important constants appear throughout the laws of electromagnetics and in transmission-line applications and should be committed to memory. The speed of light in a vacuum (and essentially in air) is

$$
\begin{aligned}
v_0 &= \frac{1}{\sqrt{\mu_0 \varepsilon_0}} \\
&= 2.99792458 \times 10^8 \\
&\cong 3 \times 10^8 \, \text{m/s}
\end{aligned}
$$

Throughout science we must deal with numbers spanning many orders of magnitude. Table 1.1 gives the common unit multipliers along with their abbreviations, which every engineer should commit to memory through frequent use.

Finally, there is an important distinction between radian frequency ω and cyclic frequency f:

$$
\omega = 2\pi f \qquad \text{rad/s}
$$

where the units of cyclic frequency f are hertz (Hz) (previously called cycles/s). For example, the cyclic frequency of commercial electric voltage and current in the United States is 60 Hz, which, in radian frequency, is 377 rad/s. The reader should never make the serious mistake of using cyclic frequency when radian frequency is required (i.e., $\omega \neq f$).

Although the SI system of units is accepted as the standard throughout the world, a few countries (including the United States) have not converted to SI units. Although there has been a strong attempt in the United States to convert to SI units, non-SI units are still in widespread use. Hence we have no choice but to learn to convert between the two systems of units. In the United States

TABLE 1.1. Unit Multipliers

Prefix	Multiplier	Symbol
giga	10^9	G
mega	10^6	M
kilo	10^3	k
centi	10^{-2}	c
milli	10^{-3}	m
micro	10^{-6}	μ
nano	10^{-9}	n
pico	10^{-12}	p

the prevalent system of units is the *English system*, where the units of length are inches (in), feet (ft), yards (yd), and statute miles (mi). The important conversion of inches to centimeters:

$$1 \text{ in} = 2.54 \text{ cm}$$

allows the conversion between units of length in the SI and English systems. The remaining units of length in the English system are:

$$1 \text{ foot(ft)} = 12 \text{ inches}$$

$$1 \text{ yard(yd)} = 3 \text{ feet}$$

$$1 \text{ mile(mi)} = 5280 \text{ ft}$$

To convert flawlessly between units that are in the different systems, simply multiply by unit ratios: for examples $1 \text{mi} = 5280 \text{ ft}$, $1 \text{ ft} = 12 \text{ in}$, $1 \text{ in} = 2.54$ cm, $1 \text{ m} = 100 \text{ cm}$, $1 \text{ km} = 1000 \text{ m}$. For example, to convert 100 miles to kilometers, we write

$$100 \text{ mi} \times \underbrace{\frac{5280 \text{ ft}}{1 \text{ mi}} \times \frac{12 \text{ in}}{1 \text{ ft}} \times \frac{2.54 \text{ cm}}{1 \text{ in}}}_{\text{English}} \times \underbrace{\frac{1 \text{ m}}{100 \text{ cm}} \times \frac{1 \text{ km}}{1000 \text{ m}}}_{\text{metric or SI}} = 160.93 \text{ km}$$

Cancellation of the unit names in this conversion avoids the improper multiplication (division) by a unity ratio when division (multiplication) should be used. There is one other dimensional unit that is used very frequently in the United States to give the dimensions of lands and PCBs: the unit of mils. One mil is one-one thousandth of an inch:

$$1000 \text{ mils} = 1 \text{ in}$$

To convert a dimension given in mils to the equivalent dimension in meters, we multiply by 2.54×10^{-5} m/mil or $1 \text{ mil} = 25.4$ micrometers (microns). For example, a common width of PCB lands is 10 mils, which is equivalent to 0.254 mm. A common thickness of printed circuit boards is 64 mils, which is equivalent to 1.63 mm. It is also common to state the per-unit-length internal inductance of a round wire in nanohenrys per inch. The exact value in SI units is 0.5×10^{-7} H/m. This is converted as

$$0.5 \times 10^{-7} \frac{H}{m} \times \frac{1 \text{ m}}{100 \text{ cm}} \times \frac{2.54 \text{ cm}}{1 \text{ in}} \times \frac{10^9 \text{ nH}}{1 \text{ H}} = 1.27 \text{ nH/in}$$

1.2   WAVES, TIME DELAY, PHASE SHIFT, WAVELENGTH, AND ELECTRICAL DIMENSIONS

In the analysis of electric circuits using Kirchhoff's voltage and current laws and lumped-circuit models, we *ignored* the connection leads attached to the lumped elements. When is this permissible? Consider the lumped-circuit element having attachment leads of total length $\mathscr{L}$ shown in Fig. 1.2. *Single-frequency sinusoidal currents* along the attachment leads are, in fact, *waves*, which can be written in terms of position z along the leads and time t as

$$i(z, t) = I\cos(\omega t - \beta z) \tag{1.1}$$

where the radian frequency ω is written in terms of cyclic frequency f as $\omega = 2\pi f$ rad/s and β is the *phase constant* in units of rad/m. (Note that the argument of the cosine must be in radians and not degrees.) To observe the movement of these current waves along the connection leads, we observe and track the movement of a point on the wave in the same way as we observe the movement of an ocean wave at the seashore. Hence the argument of the cosine in (1.1) must remain constant in order to track the movement of a point on the wave so that $\omega t - \beta z = C$, where C is a constant. Rearranging this as $z = (\omega/\beta)t - C/\beta$ and differentiating with respect to time gives the velocity

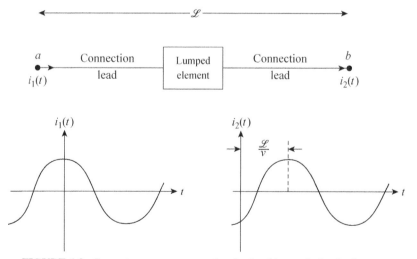

FIGURE 1.2. Current waves on connection leads of lumped-circuit elements.

of propagation of the wave as

$$v = \frac{\omega}{\beta} \quad \text{m/s} \qquad (1.2)$$

Since the argument of the cosine, $\omega t - \beta z$, in (1.1) must remain a constant in order to track the movement of a point on the wave, as time t increases so must the position z. Hence the form of the current wave in (1.1) is said to be a *forward-traveling wave* since it must be traveling in the $+z$ direction in order to keep the argument of the cosine constant for increasing time. Similarly, a *backward-traveling wave* traveling in the $-z$ direction would be of the form $i(z, t) = I\cos(\omega t + \beta z)$ since as time t increases, position z must decrease in order to keep the argument of the cosine constant and thereby track the movement of a point on the waveform. Since the current is a *traveling wave*, the current entering the leads, $i_1(z, t)$, and the current exiting the leads, $i_2(z, t)$, are separated in time by a *time delay* of

$$T_D = \frac{\mathcal{L}}{v} \quad \text{s} \qquad (1.3)$$

as illustrated in Fig. 1.2. These single-frequency waves suffer a *phase shift* of $\phi = \beta z$ radians as they propagate along the leads. Substituting (1.2) for $\beta = \omega/v$ into the equation of the wave in (1.1) gives an equivalent form of the wave as

$$i(z, t) = I\cos\left[\omega\left(t - \frac{z}{v}\right)\right] \qquad (1.4)$$

which indicates that *phase shift is equivalent to a time delay.*

Figure 1.2 plots the current waves *versus time*. Figure 1.3 plots the current wave *versus position in space at fixed times*. As we will see, the critical property of a traveling wave is its *wavelength*, denoted as λ. A wavelength *is the distance the wave must travel in order to shift its phase by 2π radians or $360°$.* Hence $\beta\lambda = 2\pi$, or

$$\lambda = \frac{2\pi}{\beta} \quad \text{m} \qquad (1.5)$$

Substituting the result in (1.2) for β in terms of the wave velocity of propagation v gives an alternative result for computing the wavelength:

$$\lambda = \frac{v}{f} \quad \text{m}$$ (1.6)

Table 1.2 gives the wavelengths of single-frequency sinusoidal waves in free space (essentially, air), where $v_0 \cong 3 \times 10^8$. (The velocities of propagation of current waves on the lands of a PCB are less than in free space, which is due to the interaction of the electric fields with the board material. Hence wavelengths on a PCB are shorter than they are in free space.) Observe that a wave of frequency 300 MHz has a wavelength of 1 m. Wavelengths scale linearly with frequency. As frequency decreases, the wavelength increases,

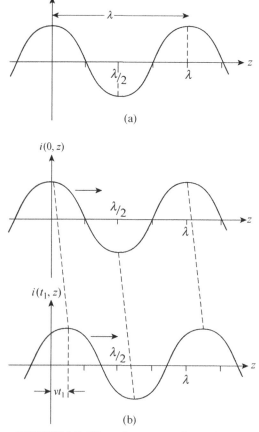

FIGURE 1.3. Waves in space, and wavelength.

TABLE 1.2. Frequencies of Sinusoidal Waves in Free Space (Air) and Their Corresponding Wavelengths

Frequency, f	Wavelength, λ
60 Hz	3107 mi (5000 km)
3 kHz	100 km
30 kHz	10 km
300 kHz	1 km
3 MHz	100 m ($\approx$ 300 ft)
30 MHz	10 m
300 MHz	**1 m** ($\approx$ 3 ft)
3 GHz	10 cm ($\approx$ 4 in)
30 GHz	1 cm
300 GHz	0.1 cm

and vice versa. For example, the wavelength of a 7 MHz wave is easily computed as

$$\lambda|_{@\,7\text{ MHz}} = \frac{300\text{ MHz}}{7\text{ MHz}} \times 1\text{ m} = 42.86\text{ m}$$

Similarly, the wavelength of a 2-GHz cell phone wave is 15 cm, which is approximately 6 in.

Now we turn to the important criterion of physical dimensions in terms of wavelengths: that is, *electrical dimensions*. To determine a physical dimension, $\mathcal{L}$, in terms of wavelengths (its electrical dimension), we write $\mathcal{L} = k\lambda$ and determine the length in wavelengths as

$$\boxed{k = \mathcal{L}/\lambda = (\mathcal{L}/v)f}$$

where we have substituted the wavelength in terms of the frequency and velocity of propagation as $\lambda = v/f$. Hence we obtain an important relation for the electrical length in terms of frequency and time delay:

$$\boxed{\begin{aligned} \frac{\mathcal{L}}{\lambda} &= f\frac{\mathcal{L}}{v} \\ &= fT_D \end{aligned}} \tag{1.7}$$

Hence a dimension is one wavelength, $\mathcal{L} = 1\lambda$, at a frequency that is the inverse of the time delay:

$$\boxed{f|_{\mathcal{L}=1\lambda} = \frac{1}{T_D}} \tag{1.8}$$

A *single-frequency sinusoidal wave* shifts phase as it travels a distance $\mathscr{L}$ of

$$
\begin{aligned}
\phi &= \beta\mathscr{L} \\
&= 2\pi\frac{\mathscr{L}}{\lambda} \qquad \text{rad} \\
&= \frac{\mathscr{L}}{\lambda} \times 360° \qquad \text{deg}
\end{aligned}
\tag{1.9}
$$

Hence if a wave travels a distance of one wavelength, $\mathscr{L} = 1\lambda$, it shifts phase by $\phi = 360°$. If the wave travels a distance of one-half wavelength, $\mathscr{L} = \frac{1}{2}\lambda$, it shifts phase by $\phi = 180°$. This can provide for cancellation: for example, when two antennas that are separated by a distance of one-half wavelength transmit the same frequency signal. Along a line containing the two antennas, the two radiating waves being of opposite phase cancel each other, giving a result of zero. This is the essential reason why antennas have "patterns" where a null is produced in one direction, whereas a maximum is produced in another direction. Phased-array radars electronically "steer" their beams using this principle rather than by rotating the antennas mechanically. Next, consider a wave that travels a distance of one-tenth of a wavelength, $\mathscr{L} = \frac{1}{10}\lambda$. The phase shift incurred in doing so is only $\phi = 36°$, and a wave that travels one-one-hundredth of a wavelength, $\mathscr{L} = \frac{1}{100}\lambda$, incurs a phase shift of $\phi = 3.6°$. Hence we say that

For any distance less than, say, $\mathscr{L} < \frac{1}{10}\lambda$, the phase shift is said to be negligible and the distance is said to be electrically short.

For electric circuits whose maximum physical dimension is electrically short, $\mathscr{L} < \frac{1}{10}\lambda$, Kirchhoff's voltage and current laws and other lumped-circuit analysis solution methods work very well. For physical dimensions that are not electrically short, Kirchhoff's laws and lumped-circuit analysis methods *give erroneous answers*! For example, consider an electric circuit that is driven by a 10-kHz sinusoidal source. The wavelength at 10 kHz is 30 km (18.641 mi)! Hence at this frequency any circuit having a maximum dimension of less than 3 km (1.86 mi) can be analyzed successfully using Kirchhoff's laws and lumped-circuit analysis methods. Electric power distribution systems operating at 60 Hz can be analyzed using Kirchhoff's laws and lumped-circuit analysis principles as long as their physical dimensions, such as the transmission-line length, are less than some 310 mi! Similarly, a circuit driven by a 1-MHz sinusoidal source can be analyzed successfully using lumped-circuit analysis methods if its maximum physical dimension is less than 30 m! On the other hand, cell phone electronic circuits operating at a frequency of around 2 GHz cannot be analyzed using lumped-circuit analysis methods unless the maximum dimension is less than around 1.5 cm, about 0.6 in! We can alternatively determine the frequency where a dimension is electrically short in terms of the time delay from (1.7):

$$f\big|_{\mathscr{L}=(1/10)\lambda} = \frac{1}{10T_D} \qquad (1.10)$$

Substituting $\lambda f = v$ into the time-delay expression in (1.3) gives the time delay as a portion of the period of the sinusoid, T:

$$\begin{aligned} T_D &= \frac{\mathscr{L}}{v} \\ &= \frac{\mathscr{L}}{\lambda}\frac{1}{f} \\ &= \frac{\mathscr{L}}{\lambda}T \end{aligned} \qquad (1.11)$$

where the period of the sinusoidal wave is $T = 1/f$. This shows that if we plot the current waves in Fig. 1.2 that enter and leave the connection leads versus time t *on the same time plot*, they will be displaced in time by a fraction of the period, $\mathscr{L}/\lambda$. If the length of the connection leads $\mathscr{L}$ is electrically short at this frequency, the two current waves will be displaced from each other *in time* by an inconsequential amount, $T/10$, and may be considered to be coincident in time. This is the reason that Kirchhoff's laws and lumped-circuit analysis methods work well only for circuits whose maximum physical dimension is "electrically small."

Waves propagated along transmission lines and radiated from antennas are of the same mathematical form as the currents on the connection leads of an element shown in (1.1). These are said to be *plane waves* where the electric and magnetic field vectors lie in a plane *transverse* or perpendicular to the direction of propagation of the wave, as shown in Fig. 1.4. These are termed *Transverse ElectroMagnetic* (TEM) *waves*.

1.3   THE TIME DOMAIN VS. THE FREQUENCY DOMAIN

Perhaps the most important concept for all engineers is that of the time and frequency domains. All engineers are used to being able to think of a problem in either domain with considerable flexibility. A problem may be more easily understood and solved in one domain than the other. For example, electric

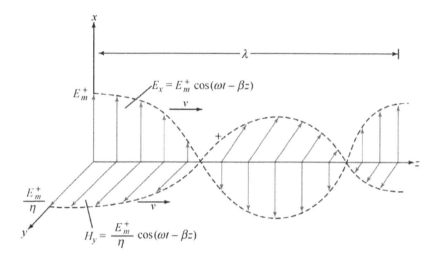

FIGURE 1.4. Electric and magnetic fields of plane waves on transmission lines and radiated by antennas.

filters are more easily designed in the frequency domain. Essentially, the time domain is the actual problem where voltages, currents, and so on, are viewed in terms of their time variations. The frequency domain views these time-domain variations in terms of their sinusoidal frequency components via the Fourier series or transform.

1.3.1   Spectra of Digital Signals

A time-domain waveform $x(t)$ is periodic if $x(t \pm mT) = x(t)$, where T is the *period* of the waveform and m is an integer. In other words, a periodic waveform repeats itself in intervals of length T. The *Fourier series* allows us to view a periodic waveform *alternatively* as being composed of the sum of a constant (dc) term and an infinite sum of sinusoids having frequencies (harmonics) that are integer multiples of the fundamental frequency, which is the inverse of the period, $f_0 = 1/T$. Hence the Fourier series allows us, alternatively, to represent a periodic waveform as being composed of sinusoidal components:

$$x(t) = c_0 + c_1\cos(\omega_0 t + \theta_1) + c_2\cos(2\omega_0 t + \theta_2) + c_3\cos(3\omega_0 t + \theta_3) + \cdots$$

$$= c_0 + \sum_{n=1}^{\infty} c_n\cos(n\omega_0 t + \theta_n)$$

$$(1.12)$$

where $\omega_0 = 2\pi f_0$. The constant (dc component) c_0 is the *average value* of the waveform over one period:

$$c_0 = \frac{1}{T}\int_0^T x(t)dt \tag{1.13a}$$

The magnitudes and angles of the sinusoidal components are computed from

$$c_n \angle \theta_n = \frac{2}{T}\int_0^T x(t)e^{-jn\omega_0 t}dt \tag{1.13b}$$

where $j = \sqrt{-1}$ and $e^{-jn\omega_0 t} = \cos n\omega_0 t - j\sin n\omega_0 t$. For a single pulse or, equivalently, as $T \to \infty$, these discrete frequency components merge into a continuous spectrum which is called the Fourier transform of the single pulse.

For the digital clock spectrum shown in Fig. 1.1, a general expression can be obtained for the magnitudes and angles of the Fourier components in (1.13), but the result is somewhat complicated and little insight is gained from it. However, if we restrict the result to trapezoidal pulses having equal rise and fall times, $\tau_r = \tau_f$ (which digital clock and data waveforms tend to approximate), we can obtain a simple and informative result. For the case of equal rise and fall times, we obtain

$$c_0 = A\frac{\tau}{T}$$

$$c_n\angle\theta_n = 2A\frac{\tau}{T}\frac{\sin(n\pi\tau/T)}{(n\pi\tau/T)}\frac{\sin(n\pi\tau_r/T)}{n\pi\tau_r/T}\angle -n\pi\frac{\tau+\tau_r}{T} \qquad \tau_r = \tau_f$$

$$\tag{1.14}$$

This result is in the form of the product of two $\sin(x)/x$ expressions, with the first depending on the ratio of the pulse width to the period, τ/T (also called the *duty cycle* of the waveform $D = \tau/T$) and the second depending on the ratio of the pulse rise and fall times to the period, τ_r/T. [The *magnitude* of the coefficient, denoted c_n, must be a positive number. Hence there may be an additional $\pm 180°$ added to the angle shown in (1.14), depending on the signs of each $\sin(x)$ term.] If, in addition to the rise and fall times being equal, the duty cyle is 50%, that is, the pulse is "on" for half the period and "off" for the other half of the period (which digital waveforms also tend to approximate), $\tau = \frac{1}{2}T$, the result for the coefficients given in (1.14) simplifies to

$$c_0 = \frac{A}{2}$$

$$c_n \angle \theta_n = A \frac{\sin(n\pi/2)}{n\pi/2} \frac{\sin(n\pi\tau_r/T)}{n\pi\tau_r/T} \angle - n\pi\left(\frac{1}{2} + \frac{\tau_r}{T}\right) \qquad \tau_r = \tau_f, \quad \tau = T/2$$

Note that the first $\sin(x)/x$ function is zero for n even, so that for *equal rise and fall times and a 50% duty cycle, the even harmonics are zero and the spectrum consists only of odd harmonics.* By replacing n/T with the smooth frequency variable f, $n/T \rightarrow f$, we obtain the *envelope* of the magnitudes of these discrete frequencies as

$$
\boxed{c_n = 2A \frac{\tau}{T} \left| \frac{\sin n\pi f \tau}{\pi f \tau} \right| \left| \frac{\sin(\pi f \tau_r)}{\pi f \tau_r} \right| \qquad \begin{array}{l} \tau_r = \tau_f \\ \dfrac{n}{T} \rightarrow f \end{array}}
\qquad (1.15)
$$

In doing so, remember that the spectral components occur only at the discrete frequencies f_0, $2f_0$, $3f_0$, ...

Observe some important properties of the $\sin(x)/x$ function:

$$\lim_{x \to 0} \frac{\sin(x)}{x} = 1$$

which relies on the property that $\sin(x) \cong x$ for small x (or using l'Hôpital's rule) and

$$\left| \frac{\sin(x)}{x} \right| \leq \begin{cases} 1 & x \leq 1 \\ \dfrac{1}{x} & x \geq 1 \end{cases}$$

The second property allows us to obtain a *bound* on the magnitudes of the c_n coefficients and relies on the fact that $|\sin(x)| \leq 1$ for all x.

A *square wave* is the trapezoidal waveform where the rise and fall times are zero:

$$c_0 = A \frac{\tau}{T}$$

$$c_n \angle \theta_n = 2A \frac{\tau}{T} \frac{\sin(n\pi\tau/T)}{n\pi\tau/T} \angle - n\pi \frac{\tau}{T} \qquad \tau_r = \tau_f = 0$$

If the duty cycle of the square wave is 50%, this result simplifies to

$$c_0 = \frac{A}{2}$$

$$c_n \angle \theta_n = \begin{cases} \dfrac{2A}{n\pi} \angle -\dfrac{\pi}{2} & n \text{ odd} \\ \\ 0 & n \text{ even} \end{cases} \qquad \tau_r = \tau_f = 0, \quad \tau = T/2$$

Figure 1.5 shows a plot of the magnitudes of the c_n coefficients for a square wave where the rise and fall times are zero, $\tau_r = \tau_f = 0$. The spectral components appear only at *discrete* frequencies, $f_0, 2f_0, 3f_0, \ldots$. The envelope is shown by a dashed line. Observe that the envelope goes to zero where the argument of $\sin \pi f \tau$ becomes a multiple of π at $f = 1/\tau, 2/\tau, \ldots$.

A more useful way of plotting the envelope of the magnitudes of the spectral coefficients is by plotting the horizontal frequency axis logarithmically and similarly plotting the magnitudes of the coefficients along the vertical axis in decibels as $|c_n|_{dB} = 20\log_{10}|c_n|$. The envelope as well as the bounds of the magnitudes of the $\sin(x)/x$ function are shown in Fig. 1.6. Observe that the actual result is bounded by 1 for $x \le 1$ and decreases at a rate of -20 dB/decade for $x \ge 1$. This rate is equivalent to a $1/x$ decrease. Also note that the magnitudes of the actual spectral components go to zero where the argument of $\sin(x)$ goes to a multiple of π or $x = \pi, 2\pi, 3\pi, \ldots$.

The amplitudes of the spectral components of a trapezoidal waveform where $\tau_r = \tau_f$ given in (1.14) are the product of two $\sin(x)/x$ functions: $\sin(x_1)/x_1 \times \sin(x_2)/x_2$. When log-log axes are used, this gives the result for the bounds on the amplitudes of the spectral coefficients shown in Fig. 1.7. Note that the bounds are constant (0 dB/decade) out to the first breakpoint of

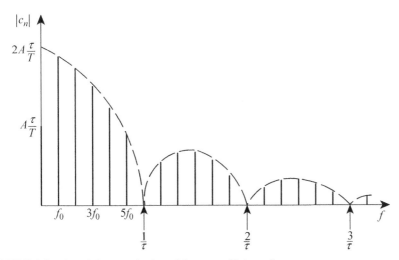

FIGURE 1.5. Plot of the magnitudes of the c_n coefficients for a square wave, $\tau_r = \tau_f = 0$.

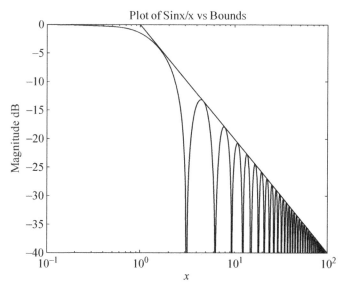

FIGURE 1.6. The envelope and bounds of the $\sin(x)/x$ function are plotted with logarithmic axes.

$f_1 = 1/\pi\tau = f_0/\pi D$, where the *duty cycle is* $D = \tau/T = \tau f_0$. Above this they decrease at a rate of $-20\,\mathrm{dB/decade}$ out to a second breakpoint of $f_2 = 1/\pi\tau_r$, and decrease at a rate of $-40\,\mathrm{dB/decade}$ above that. This plot shows the important result that *the high-frequency spectral content of the trapezoidal*

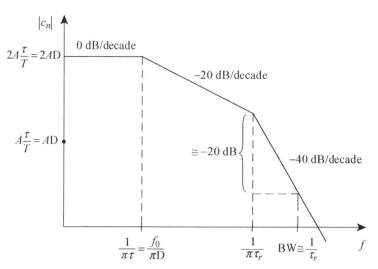

FIGURE 1.7. Bounds on the spectral coefficients of the trapezoidal pulse train for equal rise and fall times, $\tau_r = \tau_f$.

clock waveform is determined by the pulse rise and fall times. Longer rise and fall times push the second breakpoint lower in frequency, thereby reducing the high-frequency spectral content. Shorter rise and fall times push the second breakpoint higher in frequency, thereby increasing the high-frequency spectral content.

1.3.2 Bandwidth of Digital Signals

Although the Fourier series in (1.12) shows that a time-domain waveform can be thought of equivalently as being composed of the sum of sinusoidal frequency components whose frequencies are integer multiples of the fundamental frequency $f_0 = 1/T$, the sum must, ideally, include an *infinite number* of harmonic frequency components. It is, of course, not possible to include an infinite number of harmonics, so a question arises as to *how many frequency components should be included in order to give a reasonable reconstruction of the waveform.* Hence we must construct a finite-term approximation to $x(t)$ using the dc component and the first NH harmonics of the Fourier series as

$$\tilde{x}(t) = c_0 + \sum_{n=1}^{NH} c_n \cos(n\omega_0 t + \theta_n) \tag{1.16}$$

where NH represents the number of harmonics. To judge how well the finite-term approximation $\tilde{x}(t)$ in (1.16) approximates the actual $x(t)$, we examine *in time* the *pointwise errors* between the actual waveform and the finite-term approximation: $x(t) - \tilde{x}(t)$. One way of quantifying the approximation error is with the *mean-square error criterion* (MSE):

$$\text{MSE} = \frac{1}{T} \int_0^T [x(t) - \tilde{x}(t)]^2 dt \tag{1.17}$$

By squaring the pointwise errors we weight a negative pointwise error, $x(t) < \tilde{x}(t)$, and a positive pointwise error, $x(t) > \tilde{x}(t)$, equally, as we should, since either is equally detrimental. The MSE adds the pointwise errors squared over a period and averages that over the period. It can be shown that this criterion is equivalent to giving the difference in the average power in the actual waveform:

$$P_{AV} = \frac{1}{T} \int_0^T x^2(t) dt \tag{1.18}$$

and the average power in the finite-term approximation:

$$\tilde{P}_{AV} = \frac{1}{T} \int_0^T \tilde{x}^2(t)dt \tag{1.19}$$

as

$$MSE = P_{AV} - \tilde{P}_{AV} \tag{1.20}$$

For the trapezoidal clock waveform in Fig. 1.1 having equal rise and fall times, $\tau_r = \tau_f$, and a 50% duty cycle, the average power in the waveform is

$$P_{AV} = A^2 \left(\frac{1}{2} - \frac{1}{3} \frac{\tau_r}{T} \right) \tag{1.21}$$

The average power in the finite-term approximation can be shown to be

$$\tilde{P}_{AV} = c_0^2 + \sum_{n=1}^{NH} \frac{c_n^2}{2} \tag{1.22}$$

Hence the average power in the finite-term approximation is the sum of the average powers in the dc component and the NH sinusoidal harmonic components, which is Parseval's theorem. The choice of the Fourier coefficients in (1.13) causes the Fourier series to converge uniformly, meaning that adding terms successively causes the mean-square error between the actual waveform and the finite-term approximation to decrease uniformly. Successively adding harmonic components uniformly gives a better approximation to the original waveform. Note that the units of the MSE are watts (W).

The MSE criterion in (1.17) adds the squares of the pointwise errors in order to give equal weight to negative and positive pointwise errors. A more logical error criterion is to sum the *absolute values of the pointwise errors*, giving the *mean absolute error criterion* (MAE):

$$MAE = \frac{1}{T} \int_0^T |x(t) - \tilde{x}(t)| dt \tag{1.23a}$$

Substituting (1.12) and (1.16) gives an alternative expression for the MAE:

$$MAE = \frac{1}{T} \int_0^T \left| \sum_{n=NH+1}^{\infty} c_n \cos(n\omega_0 t + \theta_n) \right| dt \tag{1.23b}$$

Hence the *pointwise* error is simply the remainder terms of the Fourier series for $x(t)$ for $n > NH$ in (1.12):

$$x(t) - \tilde{x}(t) = \sum_{n=NH+1}^{\infty} c_n \cos(n\omega_0 t + \theta_n)$$

which makes sense. Since the higher-order coefficients tend to decrease with increasing n, this makes sense. Note that if the units of $x(t)$ are volts (V), the units of the MAE are also volts (V). Hence the MAE gives a more appropriate measure of the pointwise approximation error $x(t) - \tilde{x}(t)$. This error criterion again weights equally the positive and negative pointwise errors but generally cannot be integrated in closed form. This is the reason that the MSE is usually chosen over the MAE.

The important question now is: How do we define the *bandwidth* (BW) of the periodic digital clock waveform $x(t)$? A sensible criterion for choosing this is that the BW should contain the *significant spectral content of the waveform. In other words, the BW should logically be defined as the minimum number of harmonic terms required to reconstruct the original periodic waveform such that adding more harmonics gives a negligible reduction in the pointwise error, whereas using fewer harmonics gives an excessive pointwise reconstruction error.*

We investigate the answer to the question of how we choose the number of harmonics (NH) in a finite-term approximation in order to give a reasonable approximation to the actual waveform by using an example of a 1-GHz clock signal waveform (period of $T = 1$ ns) having rise and fall times of 100 ps, a 50% duty cycle, and an amplitude of $A = 5$ V. Table 1.3 shows the magnitudes and angles of the first 13 harmonics computed from (1.14), along with the wavelengths in free space, $\lambda_0 = v_0/f$, of each component. The velocities of propagation of current waves on the lands of a PCB are less than in free space, which is due to the interaction of the electric fields with the board material. Hence wavelengths on a PCB are shorter than they are in free space. Note that because of the 50% duty cycle, the even harmonics are zero. The dc component of the waveform, is the average value of the waveform; which is $A/2 = 2.5$ V.

Observe from Table 1.3 that the ninth harmonic of 9 GHz has a wavelength of 3.33 cm. To use Kirchhoff's voltage and current laws and lumped-circuit analysis principles to analyze a circuit driven by this frequency would require

TABLE 1.3. Spectral (Frequency) Components of a 5-V, 1-GHz, 50% Duty Cycle, 100-ps Rise/Fall-Time Digital Clock Signal

Harmonic	Frequency (GHz)	Wavelength, (λ_0 cm)	Level (V)	Angle (deg)
1	1	30	3.131	−108
3	3	10	0.9108	−144
5	5	6	0.4053	−180
7	7	4.29	0.1673	144
9	9	3.33	0.0387	108
11	11	2.73	0.0259	−108
13	13	2.31	0.0485	−144

that the largest dimension of the circuit be less that 3.33 mm (0.131 in)! Similarly, to analyze a circuit that is driven by the fundamental frequency of 1 GHz, whose wavelength is 30 cm, using Kirchhoff's laws and lumped-circuit analysis methods would restrict the maximum circuit dimensions to being less than 3 cm or about 1 in (2.54 cm). This shows clearly that use of lumped-circuit analysis methods to analyze a circuit having a physical dimension of, say, 1 in that is driven by this clock waveform would result in erroneous results for all but perhaps the fundamental frequency of the waveform!

Figure 1.8(a), (b), and (c) show the approximation to the clock waveform achieved by adding the dc component and the first three harmonics, the first five harmonics, and the first nine harmonics, respectively. The spectral levels and angles of the harmonic components in Table 1.3 were computed with MATLAB using the m file trapgen.m that is contained in the CD included with this book. Figure 1.8 was also generated using MATLAB m file trapgen.m.

Combining the dc term and the fundamental and third harmonics as shown in Fig. 1.8(a) gives poor convergence. Combining the dc term, the fundamental, and the third and fifth harmonics as shown in Fig. 1.8(b) while giving a better approximation still has some troublesome oscillations. Finally, combining the dc term, the fundamental, and the third, fifth, seventh, and ninth

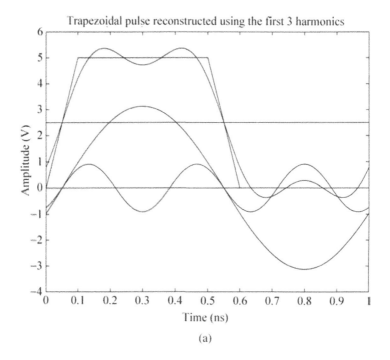

(a)

FIGURE 1.8. (a) Approximating the clock waveform with (a) the first three harmonics, (b) the first five harmonics, and (c) the first nine harmonics.

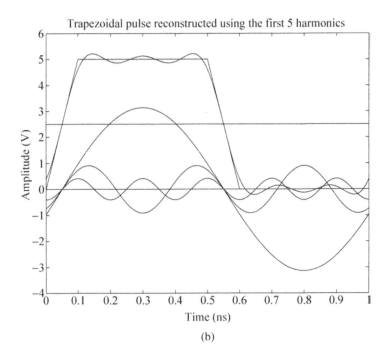

(b)

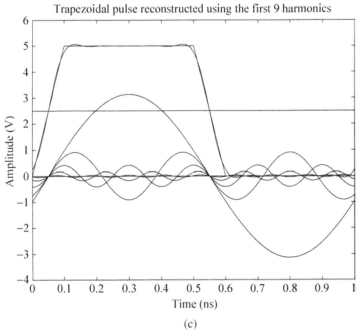

(c)

FIGURE 1.8. (*Continued*)

harmonics as shown in Fig. 1.8(c) provides an excellent reproduction of the trapezoidal waveform. It is particularly important that the approximation during the steady-state time between the rise and fall times should be A, thereby giving adequate setup and hold time intervals. Any large oscillations during this time interval may result in logic errors.

Figure 1.9(a), (b), (c), and (d) give plots of the absolute pointwise error, $|x(t) - \tilde{x}(t)|$ for NH $= 3, 5, 7$, and 9, respectively, for the 5-V 1-GHz clock waveform having a 50% duty cycle and rise and fall times of 0.1 ns. The absolute error plots in figure were plotted with MATLAB using the file plotMAE.m that is contained in the CD included with this book.

To give a more quantitative measure of how well a finite-term approximation reproduces the original waveform, we investigate the MSE and MAE error criteria for this waveform, given in (1.17) and (1.23), respectively. The MSE error criterion in (1.17) amounts to giving the difference in the average powers of the actual waveform and the finite-term approximation. The average power in the actual waveform $x(t)$ is given in (1.21) and for the example waveform is $P_{AV} = 11.667$ W. The average powers of the dc and harmonic components are given in (1.22). We can compute these from Table 1.3. The average power in the dc component is 6.25 W, and the average powers in the first 13 harmonics are 4.9 W, 0.415 W, 82.1 mW, 14 mW, 0.75 mW, 0.335 mW, and 1.18 mW. The total average power contained in the dc component and the first 13 harmonic components is 11.663 W, which is 99.97% of the total average power in the waveform! However, note that 96% of the total average power in the waveform is contained in the dc component and the first harmonic! The MSE for NH $= 1, 3, 5,$ 7, and 9 are 0.5151, 0.1003, 0.0182, 0.0042, and 0.0035 W, respectively.

The MAE error criterion in (1.23) gives a more *meaningful* quantitative measure of the approximation error $x(t) - \tilde{x}(t)$. The MAE in (1.23) cannot be integrated in closed form, so we obtain a numerical value for the MAE for the example waveform using a trapezoidal numerical integration routine. Table 1.4 gives the MAE for the 1-GHz clock waveform for various numbers of harmonics. The MAEs in the table were computed using the MATLAB m file plotMAE.m that is contained in the CD included with this book.

The results in Table 1.4 are plotted in Fig. 1.10. Note that the MAE reaches a somewhat minimum level after NH $= 9$ harmonics are used. Hence adding more harmonics gives little additional reduction in the reconstruction error, whereas using fewer than nine harmonics gives an increasingly larger reconstruction error.

How do we quantitatively determine the bandwidth of a periodic clock waveform that has different parameters than those of the example above? Recall that the logical definition of the *bandwidth* (BW) of the waveform is that the BW should be the *significant spectral content of the waveform*. In other words, the BW should be the minimum number of harmonic terms required to reconstruct the original periodic waveform such that adding more

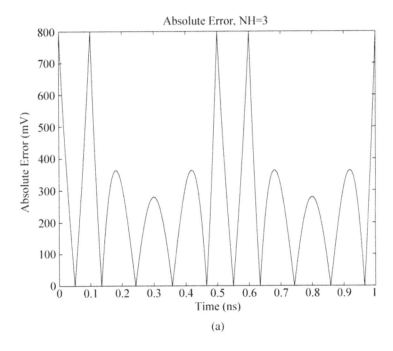

(a)

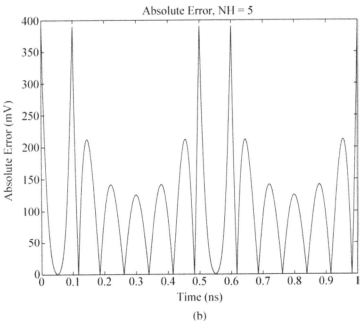

(b)

FIGURE 1.9. Plot of the pointwise absolute error for (a) NH = 3, (b) NH = 5, (c) NH = 7, and (d) NH = 9 for the 5-V 1-GHz clock waveform having a 50% duty cycle and rise and fall times of 0.1 ns.

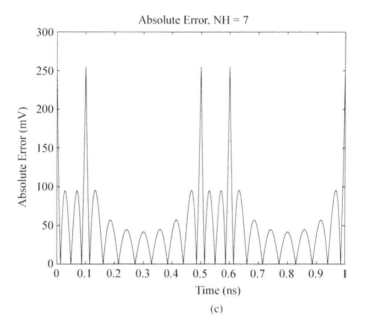

(c)

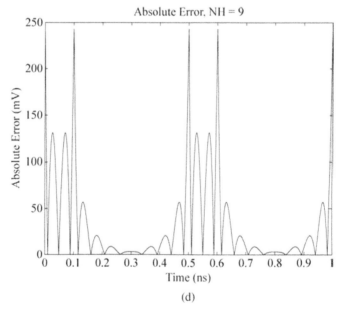

(d)

FIGURE 1.9. (*Continued*)

TABLE 1.4. Mean Absolute Error for Various Numbers of Harmonics for a 5-V, 1-GHz, 100-ps Rise/Fall Time, 50% Duty Cycle Trapezoidal Waveform

Number of Harmonics	Mean Absolute Error (mV)
1	600.3
3	265.2
5	110.1
7	50.2
9	35.0
11	39.3
13	33.0
15	22.1
17	15.5
19	12.2
21	13.9
23	12.8
25	9.9

harmonics gives a negligible reduction in the pointwise error, whereas using fewer harmonics gives an excessive pointwise reconstruction error. If we look at the plot of the bounds on the magnitude spectrum shown in Fig. 1.7, we see that above the second breakpoint, $f_2 = 1/\pi\tau_r$, the levels of the harmonics are rolling off at a rate of -40 dB/decade. If we go past this second breakpoint by

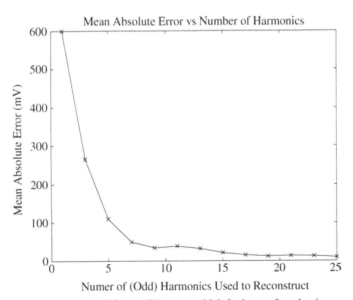

FIGURE 1.10. Plot of the MAE for a 1-GHz trapezoidal clock waveform having equal rise and fall times of $\tau_r = \tau_f = 100$ ps and a 50% duty cycle for various numbers of harmonics used to reconstruct it.

a factor of about 3 to a frequency that is the inverse of the rise and fall time, $f = 1/\tau_r$, the levels of the component at the second breakpoint will have been reduced further, by approximately 20 dB. Hence above this frequency the remaining frequency components are probably of such small magnitude that they do not provide any substantial contribution to the shape of the resulting waveform. Hence we might define the *bandwidth* of the trapezoidal clock waveform (and other data waveforms of similar shape) to be

$$\boxed{BW \cong \frac{1}{\tau_r}} \tag{1.24}$$

For the 1-GHz clock waveform above having a 50% duty cycle and 100-ps rise and fall times, by this criterion the bandwidth is 10 GHz. Hence for this waveform, the first nine harmonics contain the *significant spectral content of the waveform*. This correlates with the data in Fig. 1.10. Figure 1.11 shows a plot of the *envelope* of the spectrum of the 1-GHz waveform for $A = 1$ V and a 50% duty cycle with rise and fall times of 0.1 ns along with the bounds shown in Fig. 1.7. Remember that the actual spectral components occur at the discrete frequencies 1, 3, 5, 7, 9, ... GHz. The spectrum for $A = 5$ V can be obtained from this by adding $20 \log_{10}(5) = 13.98$ dB. Because of the 50% duty cycle, the first breakpoint in Fig. 1.7, $f_1 = f_0/\pi D = 636.6$ MHz, occurs below the fundamental frequency of 1 GHz and hence is not shown on this plot.

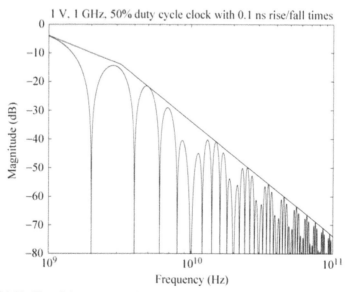

FIGURE 1.11. Plot of the spectrum of a 1-V 1-GHz clock waveform having a 50% duty cycle and rise and fall times of 0.1 ns.

The second breakpoint of $f_2 = 1/\pi\tau_r = 3.183$ GHz marks the frequency where the bounds change from -20 dB/decade to -40 dB/decade. From Table 1.3 the ninth harmonic level of 0.0387 V has a level of -28.25 dB. Subtracting $20\log_{10}(5) = 13.98$ dB gives the plotted level of -42.23 dB.

The BW criterion in (1.24) clearly does not apply to a square wave where $\tau_r = \tau_f = 0$, since it would imply that BW $=\infty$. So a corresponding BW criterion must be derived for a square wave, although an *ideal* square wave cannot be generated in practice.

The bandwidth criterion in (1.24) is not meant to be exact but is intended only to give an indication of how many frequency components should be applied to the digital system to ascertain accurately how that system would have processed the actual waveform. If the system through which the clock waveform is applied is linear (such as a transmission line having linear terminations), the principle of superposition supports this concept, as we show next.

1.3.3 Computing the Time-Domain Response of Transmission Lines Having Linear Terminations Using Fourier Methods and Superposition

Consider a linear system having an input $x(t)$ and an output of $y(t)$, as shown in Fig. 1.12. A linear system is one for which the principle of superposition applies. In other words, the system is *linear* if $x_1(t) \rightarrow y_1(t)$ and $x_2(t) \rightarrow y_2(t)$; then (1) $(x_1(t) + x_2(t)) \rightarrow (y_1(t) + y_2(t))$ and (2) $kx(t) \rightarrow ky(t)$.

We can decompose a periodic waveform into its Fourier components according to

$$x(t) = c_0 + c_1\cos(\omega_0 t + \theta_1) + c_2\cos(2\omega_0 t + \theta_2) + c_3\cos(3\omega_0 t + \theta_3) + \cdots$$

$$(1.12)$$

If we pass each component of the waveform through the linear system individually and sum, *in time*, the responses to these components, we will have determined, indirectly, the response of the system to the entire waveform being passed through it, as indicated in Fig. 1.13. It is important to note that

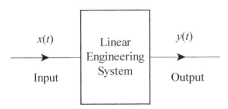

FIGURE 1.12. Single-input, single-output linear system.

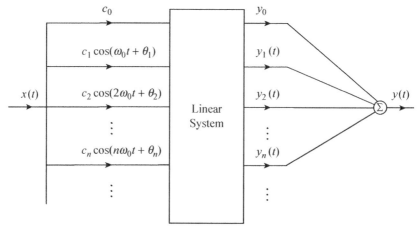

FIGURE 1.13. Using superposition to determine the (steady-state) response of a linear system to a waveform by passing the individual Fourier components through the system and summing their responses at the output.

this technique gives only the *steady-state response*. In other words, any transient portion of the response to the original signal is not included.

As an example of this powerful technique, consider an RC circuit that is driven by a periodic square-wave voltage source as shown in Fig. 1.14. The square wave has an amplitude of 1 V, a period of 2 s, and a pulse width of 1 s (50% duty cycle). The RC circuit consisting of the series connection of $R = 1\,\Omega$ and $C = 1$ F has a time constant of $RC = 1$ s, and the voltage across the capacitor is the desired output voltage of this linear "system." The nodes of the circuit are numbered in preparation for using the SPICE circuit analysis program (or the personal computer version, PSPICE) to analyze it and plot the exact solution. The Fourier series of the input, $v_S(t)$, using only the first seven harmonics, is ($\omega_0 = 2\pi/T = \pi$)

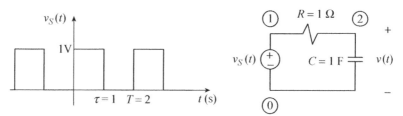

FIGURE 1.14. Example of using superposition of the Fourier components of a signal in obtaining the (steady-state) response to that signal.

$$v_S(t) = c_0 + c_1\cos(\omega_0 t + \theta_1) + c_3\cos(3\omega_0 t + \theta_3) + c_5\cos(5\omega_0 t + \theta_5)$$
$$+ c_7\cos(7\omega_0 t + \theta_7)$$
$$= \frac{1}{2} + \frac{2}{\pi}\cos(\pi t - 90°) + \frac{2}{3\pi}\cos(3\pi t - 90°) + \frac{2}{5\pi}\cos(5\pi t - 90°)$$
$$+ \frac{2}{7\pi}\cos(7\pi t - 90°)$$

The phasor (sinusoidal steady-state) transfer function of this linear system is

$$\hat{H}(jn\omega_0) = \frac{\hat{V}}{\hat{V}_S}$$
$$= \frac{1}{1 + jn\omega_0 RC}$$
$$= \frac{1}{1 + jn\pi}$$
$$= \frac{1}{\sqrt{1 + (n\pi)^2}} \angle -\tan^{-1}n\pi$$
$$= H_n \angle \phi_n$$

The phasor (sinusoidal steady state) voltages and currents will be denoted with carets and are complex-valued having a magnitude and an angle: $\hat{V} = V \angle \theta_V$ and $\hat{I} = I \angle \theta_I$. The output of this "linear system" is the voltage across the capacitor, $v(t)$, whose Fourier coefficients are obtained as $c_n H_n \angle (\theta_n + \phi_n)$, giving the Fourier series of the time-domain output waveform as

$$v(t) = 0.5 + 0.1931\cos(\pi t - 162.34°) + 0.0224\cos(3\pi t - 173.94°)$$
$$+ 0.0081\cos(5\pi t - 176.36°) + 0.0041\cos(7\pi t - 177.4°)$$

Figure 1.15 shows the approximation to the output waveform for $v(t)$ obtained by *summing in time* the steady-state responses to only the dc component and the first seven harmonics of $v_S(t)$.

The exact result for $v(t)$ is obtained with PSPICE and shown in Fig. 1.16. Note that there is an initial transient part of the solution over the first 2 or 3 s due to the capacitor being charged up to its steady-state voltage. These results make sense because as the square-wave transitions to 1 V, the voltage across the capacitor increases according to $1 - e^{-t/RC}$. Since the time constant is $RC = 1$ s, the voltage has not reached steady state (which requires about five time constants to have elapsed) when the square wave turns off at $t = 1$s. Then the capacitor voltage begins to discharge. But when the square wave turns on

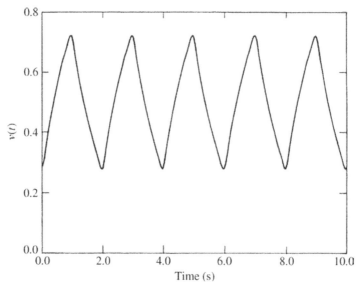

FIGURE 1.15. Voltage waveform across the capacitor of Fig. 1.14 obtained by adding the (steady-state) responses of the dc component and the first seven harmonics of the Fourier series of the square wave.

again at $t = 2$ s, the capacitor has not fully discharged and begins recharging. This process and the resulting output voltage waveform repeats with a period of 2 s. The transitions in the exact waveform of the output voltage in Fig. 1.16 are sharper than the corresponding transitions in the approximate waveform in Fig. 1.15 obtained by summing the responses to the first seven harmonics of the Fourier series of the input waveform. This is due to neglecting the

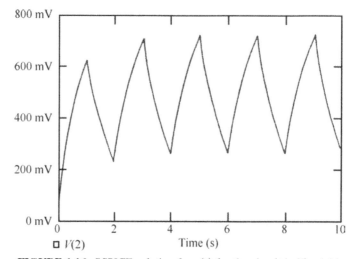

FIGURE 1.16. PSPICE solution for $v(t)$ for the circuit in Fig. 1.14.

responses to the high-frequency components of the input waveform and is a general property.

This example shows a powerful method of using superposition and the Fourier series indirectly to determine the response of a *linear* system to a complicated time-domain waveform. Quite often it is simpler to take this route rather than confronting the solution to the problem directly in the time domain by solving the differential equation relating $v(t)$ to $v_S(t)$. We will find this technique to be quite useful in determining the time-domain response of a transmission line and at the same time including frequency-dependent line losses that would make a direct time-domain solution very tedious and involved.

1.4 THE BASIC TRANSMISSION-LINE PROBLEM

A transmission line connects a *source* to a *load* as shown in Fig. 1.17(a). The objective will be to determine the time-domain response waveform of the output voltage of the line, $V_L(t)$, given the termination impedances, R_S and R_L, the source voltage waveform $V_S(t)$, and the properties of the transmission line. If the source and termination impedances are linear, we may, alternatively, view the transmission-line problem as a linear system having an input $V_S(t)$ and an output $V_L(t)$ by embedding the terminations in one system, as shown in

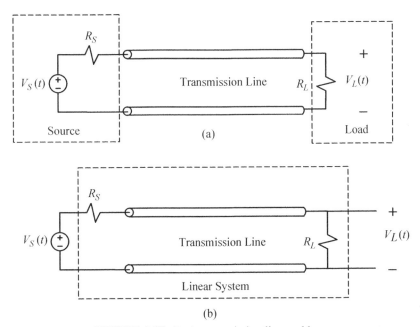

FIGURE 1.17. Basic transmission-line problem.

Fig. 1.17(b). Hence the time-domain solution for $V_L(t)$ can be obtained in an approximate fashion using superposition and summing, *in time*, the responses to the components of the Fourier series for $V_S(t)$. We will develop methods for direct sketching of the waveform of $V_L(t)$ by hand when the termination impedances are linear resistors. We will use PSPICE to solve this problem when the termination impedances are linear capacitors or inductors or a combination of these elements.

The essential questions to be answered in this book are whether the two parallel conductors that interconnect the source and the load and comprise the transmission line have any significant effect on the signal transmitted to the load, and if so, how we calculate that effect. Lumped-circuit analysis principles would suggest that the line conductors can be ignored and the load voltage can be computed by voltage division as

$$V_L(t) = \frac{R_L}{R_S + R_L} V_S(t)$$

But this result assumes that the length of the transmission line, $\mathscr{L}$, is very short, electrically (i.e. $\mathscr{L} \ll \lambda$). However, if the line length is not electrically short at *all* the significant frequencies of the source voltage (within its bandwidth), the transmission line cannot be ignored and will have the possibility of affecting adversely the signal transmitted to the load.

1.4.1   Two-Conductor Transmission Lines and Signal Integrity

Signal integrity means that the two interconnect conductors (the interconnect line) connecting two electronic modules should not appreciably affect the signal transmitted along the interconnect line to its load. In other words, for the digital modules connected by these conductors to perform reliably, we expect (hope!) that the interconnect conductors have no appreciable effect on those transmitted signals other than imposing the inevitable time delay. We study this in Part I.

Figure 1.18 is an example of an interconnecting set of conductors (lands on a PCB) causing severe logic errors, resulting in poor signal integrity. Two CMOS inverters (buffers) are connected by 2 in of lands ($\mathscr{L} = 2\,\text{in} = 0.0508\,\text{m}$) on a PCB. The output of the left inverter is shown as a Thévenin equivalent circuit having a low source resistance of $10\,\Omega$. This is fairly typical of CMOS devices except that the output resistance is somewhat nonlinear. The load on the line is the input to the other CMOS inverter, which is represented as a 5-pF capacitor, which is also typical of the input to CMOS devices. We are interested in determining the voltage at the output of the interconnect line, $V_L(t)$, which is the voltage at the input to the second CMOS inverter.

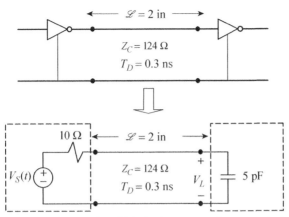

FIGURE 1.18. Example of a case in which the interconnecting transmission line can have severe effects on the signals transmitted.

For the cross-sectional dimensions of the line shown in Fig. 1.19, the characteristic impedance is $Z_C = 124\,\Omega$, and the velocity of propagation is $v = 1.7 \times 10^8$ m/s. These computations are explained later in the book. This gives a one-way time delay of

$$T_D = \frac{\mathcal{L}}{v} = \frac{0.0508\text{ m}}{1.7 \times 10^8\text{ m/s}} = 0.3\text{ ns}$$

The source voltage is a 5 V, 50-MHz ($T = 20$ ns) clock waveform having a 50% duty cycle and rise and fall times of 0.5 ns as shown in Fig. 1.20. The bandwidth of this waveform is

$$\text{BW} = \frac{1}{\tau_r} = \frac{1}{0.5 \times 10^{-9}} = 2\text{ GHz}$$

The line length of 2 in is electrically small at

$$f|_{\mathcal{L}=(1/10)\lambda} = \frac{1}{10 T_D} = \frac{1}{3 \times 10^{-9}} = 333\text{ MHz}$$

The line length is electrically short for only the first seven harmonics. However, the line is *not electrically short* for a substantial portion of the

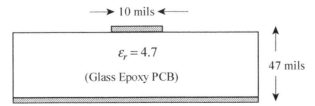

FIGURE 1.19. Cross section of the PCB in Fig. 1.18.

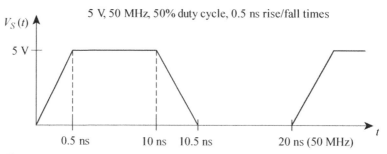

FIGURE 1.20. Source voltage waveform for the problem shown in Fig. 1.18.

spectrum of the input waveform $V_S(t)$ (from 333 MHz to 2 GHz, or 33 harmonics). Hence the interconnect line should be modeled using the transmission-line model in order to determine the load voltage $V_L(t)$ correctly.

Figure 1.21(a) shows the *ideal* load voltage, $V_L(t)$, but separated from the source voltage, $V_S(t)$, by the inevitable one-way time delay of $T_D = 0.3$ ns. This represents the ideal signal integrity solution that we wish to achieve in order for the system to work properly. Figure 1.21(b) shows the actual response for the load voltage computed with PSPICE using the exact transmission-line model contained in PSPICE. Typical thresholds for CMOS circuits are around halfway between the logic 1 and logic 0 levels, which in this case are 5 V and 0 V. Observe that there is severe "ringing" in the response and the response drops below the 2.5 V high level and rises above the 2.5 V low level thereby producing false logic triggering. Hence signal integrity is not achieved here.

If we examine this problem in the frequency domain, we obtain additional insight into why the load voltage has such severe ringing. Table 1.5 shows the Fourier components for the 5-V 50-MHz source voltage $V_S(t)$, which has a 50% duty cycle waveform and rise and fall times of 0.5 ns. Figure 1.22 shows the magnitude and angle of the (phasor) frequency response, $\hat{V}_L/\hat{V}_S$ from 1 MHz to 10 GHz, which is also computed using PSPICE and representing the interconnect line with the transmission-line model.

Observe that there is a peak in the magnitude of the frequency response of the transfer function in Fig. 1.22(a) occurring at 340 MHz of level 17.45 dB, which is also the frequency of the ringing in Fig. 1.21. This peak is due to the 5-pF load capacitor. The remaining resonances are due to the transmission line, which is one wavelength at $f|_{\mathscr{L}=1\lambda} = 1/T_D = 3.333$ GHz and occur at multiples of a half wavelength. Hence the load capacitance causes a peak in the frequency response of 17.45 dB (a ratio of 7.46) at 340 MHz, which multiplies the level of the seventh harmonic of 350 MHz (0.4322 V) by a factor of about 7.46, thereby enhancing the ringing at that frequency on the time-domain waveform. Observe that the frequency response above some 500 MHz (where

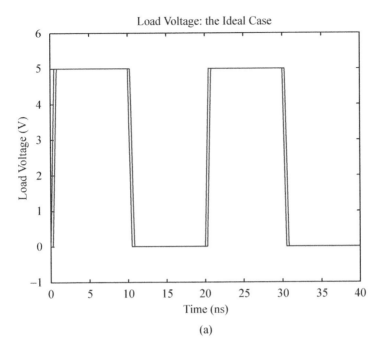

(a)

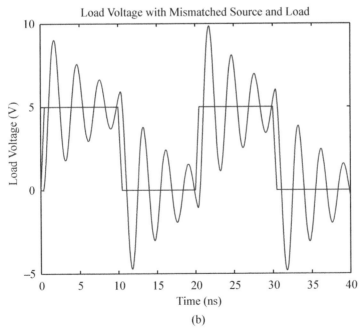

(b)

FIGURE 1.21. Load voltage, $V_L(t)$, compared to the source voltage, $V_S(t)$, for (a) the *ideal* case where signal integrity is achieved, and (b) the *actual* case indicating logic errors and false switching of the terminal inverter.

TABLE 1.5. Spectral (Frequency) Components of a 5-V 50-MHz, 50% Duty Cycle, 0.5-ns Rise/Fall-Time Digital Clock Signal

Harmonic	Frequency (MHz)	Level (V)	Angle (deg)
1	50	3.1798	−94.5
3	150	1.0512	−103.5
5	250	0.6204	−112.5
7	350	0.4322	−121.5
9	450	0.325	−130.5
11	550	0.2547	−139.5
13	650	0.2045	−148.5

the interconnect line is electrically long) decreases at a rate of − 20 dB/decade, which is due to the load capacitor. In this case, the higher-frequency harmonics of the source voltage which are rolling off in amplitude at a rate of − 40dB/decade above $f_2 = 1/\pi\tau_r = 636$ MHz (see Fig. 1.7) are rolling off in the load voltage $V_L(t)$ at a rate of − 60 dB/decade at frequencies where the line is electrically long and hence are not of much consequence. Hence a lumped-circuit model of the line shown in Fig. 1.23 may suffice in this case.

The construction of this lumped equivalent circuit is explained later in the book. We can calculate the ringing frequency approximately from this lumped equivalent circuit of the interconnect line as

$$f_{\text{ringing}} = \frac{1}{2\pi\sqrt{LC}}$$
$$= \frac{1}{2\pi\sqrt{(37.05 \text{ nH})(1.205 \text{ pF} + 5 \text{ pF})}}$$
$$= 332 \text{ MHz}$$

which is approximately the ringing frequency in Fig. 1.21. Although the lumped equivalent circuit in Fig. 1.23 is an approximate representation of the line, an insight such as this is one of the advantages of using it to approximately represent the line.

Figure 1.24 shows a comparison of the time-domain results obtained with the transmission-line model and the lumped-circuit model of Fig. 1.23. The predictions of the lumped-circuit model in Fig. 1.23 correlate reasonably well with those of the transmission-line model, although there is a slight time shift between them.

The *magnitudes* of the frequency-domain transfer functions computed by the transmission-line model and the lumped-circuit model of Fig. 1.23 are compared in Fig. 1.25. Note that the predictions of the lumped-circuit model of the line in Fig. 1.23 compare well with those of the transmission-line model

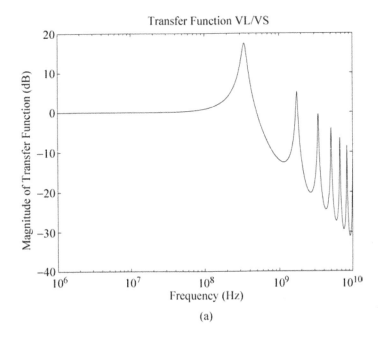

(a)

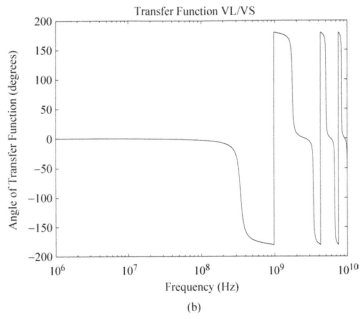

(b)

FIGURE 1.22. (a) Magnitude of the transfer function, $\hat{V}_L/\hat{V}_S$, and (b) angle $\angle\frac{\hat{V}_L}{\hat{V}_S}$ of the linear system in Fig. 1.18.

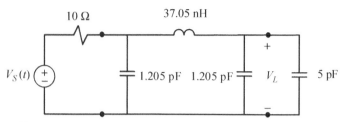

FIGURE 1.23. Lumped equivalent circuit of the interconnect line in Fig. 1.18.

for frequencies where the line is electrically short (i.e., below around 350 MHz). If we increase the rise and fall times to 5 ns and keep all other parameters unchanged, the Fourier coefficients of $V_S(t)$ are as shown in Table 1.6. Observe that the magnitudes of the Fourier coefficients are reduced considerably from those in Table 1.5, where the rise and fall times were 0.5 ns. For rise and fall times of 5 ns, the bandwidth of $V_S(t)$ is BW $= 1/\tau_r = 200$ MHz. Hence the majority of the significant frequency components fall below the resonance of 340 MHz shown in the frequency-domain transfer function in Figs. 1.22 and 1.25, and the ringing should be reduced substantially.

Figure 1.26 shows the load voltage for a 5-ns rise and fall time, with all other parameters remaining the same. Note that the ringing is reduced substantially. This is due to the fact that the magnitude of the seventh harmonic (350 MHz) is reduced from 0.4322 V for $\tau_r = 0.5$ ns to 0.0585 V for $\tau_r = 5$ ns, a factor of

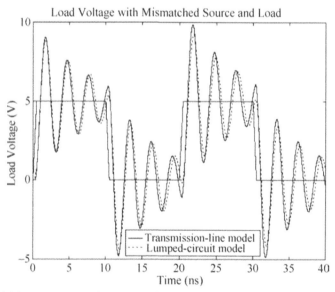

FIGURE 1.24. Comparison of the results for the load voltage for the problem of Fig. 1.18 using the transmission-line model and a lumped-circuit model of the interconnect line.

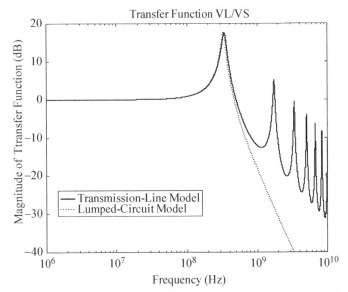

FIGURE 1.25. Predictions of the magnitude of the frequency-domain transfer function, $|\hat{V}_L/\hat{V}_S|$, of the problem in Fig. 1.18 via the transmission-line model and the lumped-circuit model of Fig. 1.23.

7.39. Figure 1.27 shows a comparison of the predictions of the transmission-line model and the lumped-circuit model of Fig. 1.23 for the case of 5-ns rise and fall times, confirming the adequacy of both models for this case.

> The key to achieving signal integrity is the relationship between the *bandwidth* (BW) *of the signal* that is carried by the interconnect conductors and the *bandwidth of the interconnect transmission line.* These are two different bandwidths.

If the frequency domain transfer function is constant (magnitude and phase) from $f = 0$ to $f =$ BW, then the load and source voltage will be identical in

TABLE 1.6. Spectral (Frequency) Components of a 5-V, 50-MHz, 50% Duty Cycle, 5-ns Rise/Fall-Time Digital Clock Signal

Harmonic	Frequency (MHz)	Level (V)	Angle (deg)
1	50	2.8658	−135
3	150	0.3184	135
5	250	0.1146	−135
7	350	0.0585	135
9	450	0.0354	−135
11	550	0.0237	135
13	650	0.0170	−135

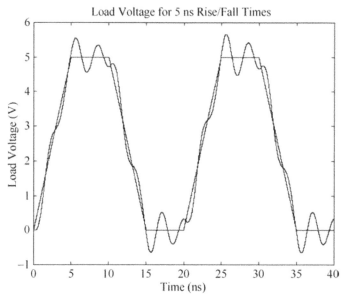

FIGURE 1.26. Time-domain load voltage for the circuit in Fig. 1.18 with the rise and fall times of $V_S(t)$ increased to 5 ns.

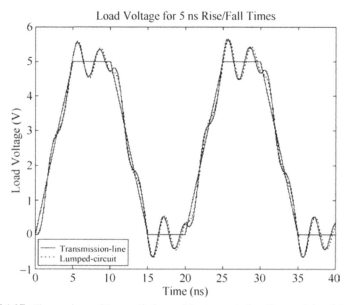

FIGURE 1.27. Comparison of the predictions of the transmission-line model and the lumped-circuit model of Fig. 1.23 for the case of 5-ns rise and fall times.

shape, and signal integrity will be achieved. For example, see the transfer function in Fig. 1.22. Essentially, the transfer function is unity from dc to around 100 MHz. If the load capacitor is replaced with a resistor, R_L, then the transfer function will be, by voltage division, at these lower frequencies, $\frac{R_L}{R_S + R_L}$, which gives the load voltage as $V_L(t) = \frac{R_L}{R_S + R_L} V_S(t)$, which also achieves signal integrity.

1.4.2 Multiconductor Transmission Lines and Crosstalk

Crosstalk is the unintended coupling of a signal on a pair of conductors onto an adjacent pair of conductors, thereby possibly causing interference in the electronic devices interconnected by the adjacent pair of conductors. This has become a serious design problem in today's high-speed digital systems. We study crosstalk in Part II.

We illustrate this problem of crosstalk by adding to the PCB of the previous problem a second land of width 10 mils. The two lands are separated edge to edge by 10 mils, as shown in Fig. 1.28. The total length of the lands is again $\mathscr{L} = 2$ in $= 0.0508$ m. The lands are terminated as shown in Fig. 1.28 with 50-Ω resistors simulating the source impedance of a commercial pulse generator and the input impedances of an oscilloscope that may be used to measure the resulting voltages at the ends of the lands. Each land with the ground plane (beneath the substrate) forms a circuit. The *generator circuit* is driven at the left end by a pulse source having a source impedance of 50 Ω and is terminated by 50 Ω at the right end. The voltage source, $V_S(t)$, again simulates a 5-V trapezoidal waveform of frequency 50 MHz and a 50% duty cycle as shown in Fig. 1.20 but having rise and fall times of 0.1 ns. The *receptor circuit* is terminated at both ends by 50 Ω resistors. The *near-end crosstalk voltage*, V_{NE}, is at the left end of the receptor circuit, and the *far-end*

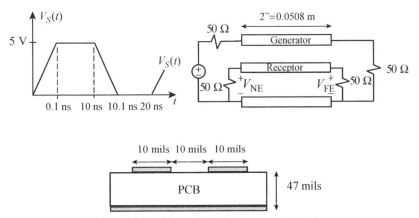

FIGURE 1.28. Illustration of the crosstalk problem.

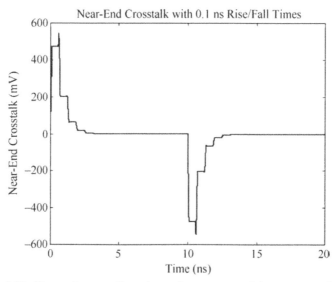

FIGURE 1.29. Near-end crosstalk voltage for $\tau_r = \tau_f = 0.1$ ns computed using the transmission-line model.

crosstalk voltage, V_{FE}, is at the right end of the receptor circuit. The following predictions of the near-end and far-end crosstalk voltages are those of the transmission-line model and are obtained with PSPICE using a crosstalk subcircuit model that we derive in Part II.

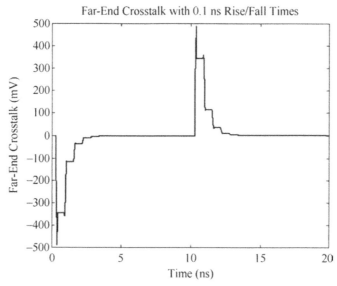

FIGURE 1.30. Far-end crosstalk voltage for $\tau_r = \tau_f = 0.1$ ns computed using the transmission-line model.

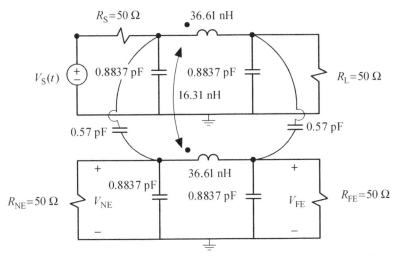

FIGURE 1.31. Lumped equivalent-circuit model of the problem in Fig. 1.28.

The bandwidth of the signal for a rise and fall time of 0.1 ns is BW $\cong$ $1/\tau_r = 10$ GHz. The one-way time delay is approximately $T_D \cong (\mathcal{L}/v_0)$ $\sqrt{\varepsilon_r'} = 0.286$ ns, where $\varepsilon_r' \cong (1 + 4.7)/2 = 2.85$ is an "effective" relative permittivity as though the surrounding medium were homogeneous and represents the average of the relative permittivity of air, $\varepsilon_r = 1$, and the relative permittivity of the PCB board, $\varepsilon_r = 4.7$. This is discussed in more depth in Part II. Hence the line is electrically short for frequencies below about

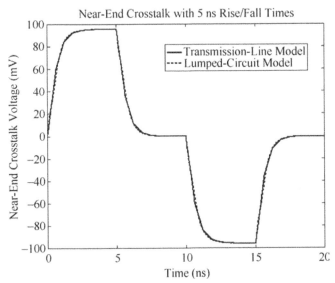

FIGURE 1.32. Near-end crosstalk voltage for $\tau_r = \tau_f = 5$ ns computed using the transmission-line and lumped-circuit models of Fig. 1.31.

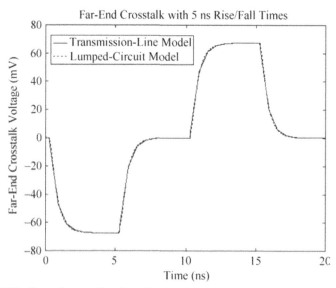

FIGURE 1.33. Far-end crosstalk voltage for $\tau_r = \tau_f = 5$ ns computed using the transmission-line and lumped-circuit models of Fig. 1.31.

$1/10T_D = 350$ MHz. Hence, for rise and fall times of 0.1 ns, there are a substantial number of spectral components of $V_S(t)$, where the transmission line is electrically long. Therefore, for these short rise and fall times we cannot model the line with a lumped-circuit model and must use the distributed-

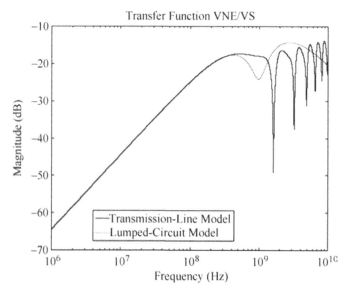

FIGURE 1.34. Comparison of the predictions of the magnitudes of the frequency-domain transfer function $\left| \hat{V}_{NE}/\hat{V}_S \right|$ by the transmission-line and lumped-circuit models of Fig. 1.31.

parameter transmission-line model. Figure 1.29 shows the near-end crosstalk voltages, and Fig. 1.30 shows the far-end crosstalk voltages computed using the exact PSPICE subcircuit model that we discuss in Part II. Observe that these crosstalk voltages occur during rise and fall times of $V_S(t)$. In Part II we find this to be a general result.

If we slow the rise/fall times of $V_S(t)$ to 5 ns, this signal will have a bandwidth of only BW $= 1/\tau_r = 200$ MHz. Since the line is electrically short for frequencies that are less than $1/10T_D = 350$ MHz, a lumped-circuit model of the line shown in Fig. 1.31 should give accurate predictions of the crosstalk voltages. The exact predictions of the transmission-line model and the predictions of the lumped-circuit model in Fig. 1.31 are compared in Figs. 1.32 and 1.33. Observe that the exact transmission-line model and the lumped-circuit model of Fig. 1.31 give virtually identical predictions of the crosstalk voltages, as expected. Note further that the crosstalk voltages appear as rectangular pulses occurring during the rise and fall times of $V_S(t)$. We confirm this observation in Part II and determine the values of these crosstalk pulses using a much simpler model for electrically short lines.

Plots of the magnitude of the frequency-domain transfer functions $\left|\hat{V}_{NE}/\hat{V}_S\right|$ in Fig. 1.34 and $\left|\hat{V}_{FE}/\hat{V}_S\right|$ in Fig. 1.35 show, as expected, that the lumped-circuit model of Fig. 1.31 and the transmission-line model give virtually identical predictions for frequencies where the line is electrically short (i.e., below around 350 MHz). Hence the close predictions of the two models for $\tau_r = \tau_f = 5$ ns in Figs 1.32 and 1.33 are to be expected.

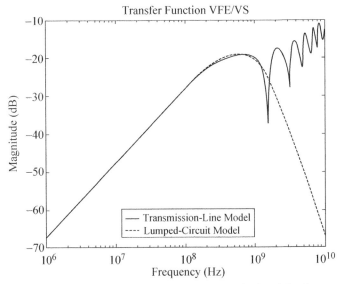

FIGURE 1.35. Comparison of the predictions of the magnitude of the frequency-domain transfer function $\left|\hat{V}_{FE}/\hat{V}_S\right|$ by the transmission-line and lumped-circuit models of Fig. 1.31.

PROBLEMS

1.1 Express the following values of resistance, capacitance, and inductance in terms of the multipliers in Table 1.1.

(a) $25 \times 10^4 \, \Omega$ [250 kΩ]

(b) $0.035 \times 10^4 \, \Omega$ [350 Ω]

(c) 0.00045 F [450 μF]

(d) 0.003×10^{-7} F [0.3nF]

(e) 0.005×10^{-2} H [50 μH]

1.2 Convert the following dimensions to those indicated.

(a) 30 mi to km [48.3 km]

(b) 1 ft to mils [12,000 mils]

(c) 100 yd (length of a U.S. football field) to m [91.44 m]

(d) 5 mm to mils [196.85 mils]

(e) 20 μm to mils [0.7874 mil]

(f) 880 yd (race distance) to m [804.67 m]

1.3 A sinusoidal current wave is described below. Determine the velocity of propagation and the wavelength. If the wave travels a distance d, determine the time delay and phase shift. Determine the frequency where the distance is one wavelength and the distance in terms of a wavelength.

(a) $i(t,z) = I_0 \sin(2\pi \times 10^6 t - 2.2 \times 10^{-2} z)$, $d = 3$ km

$[v = 2.856 \times 10^8$ m/s, $\lambda = 285.6$ m, $T_D = 10.5 \, \mu$s, ϕ
$= 3781.5° = 181.5°$, $f|_{d=1\lambda} = 95.2381$ kHz, $d = 10.5\lambda]$

(b) $i(t,z) = I_0 \sin(6\pi \times 10^9 t - 75.4z)$, $d = 4$ in

$[v = 2.5 \times 10^8$ m/s, $\lambda = 83.3$ mm, T_D
$= 0.41$ ns, $\phi = 438.92° = 78.92°$, $f|_{d=1\lambda} = 2.461$ GHz, $d = 1.219\lambda]$

(c) $i(t,z) = I_0 \sin(30\pi \times 10^7 t - 3.15z)$, $d = 20$ ft

$[v = 2.99 \times 10^8$ m/s, $\lambda = 1.995$ m, $T_D = 20.4$ ns, ϕ
$= 1100.2° = 20.2°$, $f|_{d=1\lambda} = 49.08$ MHz, $d = 3.06\lambda]$

(d) $i(t,z) = I_0 \sin(6\pi \times 10^3 t - 0.126 \times 10^{-3} z)$, $d = 50$ mil

$[v = 1.5 \times 10^8$ m/s, $\lambda = 50$ km, $T_D = 0.54$ ms, $\phi = 580.9°$
$= 220.9°$, $f|_{d=1\lambda} = 1859$ kHz, $d = 1.61\lambda]$

1.4 Determine the wavelength at the following frequencies in SI and in English units.

(a) Loran C long-range navigation at 90 Hz [3,333.3 km, 2,071.2 mil]

(b) Submarine communication at 1 kHz [300 km, 186.41 mil]

(c) Automatic direction finder in aircraft at 350 kHz [857.14 m, 0.533 mil]

(d) AM radio transmission at 1.2 MHz [250 m,820.2 ft]

(e) Amateur radio at 35 MHz [8.57 m, 28.12 ft]

(f) FM radio transmission at 110 MHz [2.73 m, 8.95 ft]

(g) Instrument landing system at 335 MHz [89.55 cm, 2.94 ft]

(h) Satellite at 6 GHz [5 cm, 1.97 in]

(i) Remote sensing at 45 GHz [6.67 mm, 262.5 mils]

1.5 Determine the following physical dimensions in wavelengths (i.e., their electrical dimension).

(a) A 50-mil length of a 60-Hz power transmission line [$1/62\lambda$]

(b) A 500-ft AM broadcast antenna broadcasting at 500 kHz [0.254λ]

(c) A 4.5-ft FM broadcast antenna broadcasting at 110 MHz [0.5λ]

(d) A 2-in land on a printed circuit board (assume a velocity of propagation of 1.5×10^8 m/s) at 2 GHz [0.677λ]

1.6 Determine the dc and first seven components of a 200-MHz clock waveform having an amplitude of 5 V, a 50% duty cycle, and rise and fall times of 0.3 ns. Determine the average power of the actual waveform and the MSE using the constant and the first seven harmonics. [2.5 V, $3.1643\angle - 100.8°$ V, $1.0054\angle - 122.4°$ V, $0.5465\angle - 144°$ V, $0.3338\angle - 165.6°$ V, 12 W, 0.0332 W]

1.7 Determine the dc and first seven components of a 70-MHz clock waveform having an amplitude of 5 V, a 50% duty cycle, and rise and fall times of 3 ns. Determine the average power of the actual waveform and the MSE using the constant and the first seven harmonics. [2.5 V, $2.9572\angle - 127.8°$ V, $0.492\angle156.6°$ V, $0.0302\angle - 99°$ V, 0.098 $\angle - 174.6°$ V, 10.75 W, 0.0013 W]

1.8 Determine the dc and first seven components of a 600-MHz clock waveform having an amplitude of 3 V, a 50% duty cycle, and rise and fall times of 0.3 ns. Determine the average power of the actual waveform and the MSE using the constant and the first seven harmonics. [1.5 V, $1.8097\angle - 122.4°$ V, $0.3723\angle172.8°$ V, $0.0417\angle108°$ V, 0.0502 $\angle - 136.8°$ V, 3.96 W, 0.0011 W]

1.9 Compute the values of the bounds on the magnitudes of the coefficients for the waveform in Problem 1.6 using the asymptotic plot in Fig. 1.7, and compare to the exact values. [3.1831 V, 1.06103 V, 0.63662 V, 0.34463 V]

1.10 Compute the values of the bounds on the magnitudes of the coefficients for the waveform in Problem 1.7 using the asymptotic plot in Fig. 1.7 and compare to the exact values. [3.1831 V, 0.5361 V, 0.193 V, 0.0985 V]

1.11 Compute the values of the bounds on the magnitudes of the coefficients for the waveform in Problem 1.8 using the asymptotic plot in Fig. 1.7 and compare to the exact values. [1.90986 V, 0.37526 V, 0.135 V, 0.06893 V]

1.12 Determine the bandwidths of the clock waveforms in Problems 1.6, 1.7, and 1.8. [3.33 GHz, 333.3 MHz, 3.33 GHz]

1.13 Plot the exact and approximate waveforms using the dc component and first seven harmonics for the clock waveform in Problem 1.6 using, for example, MATLAB.

1.14 Plot the exact and approximate waveforms using the dc component and first seven harmonics for the clock waveform in Problem 1.7 using, for example, MATLAB.

1.15 Plot the exact and approximate waveforms using the dc component and first seven harmonics for the clock waveform in Problem 1.8 using, for example, MATLAB.

PART I

TWO-CONDUCTOR LINES
AND SIGNAL INTEGRITY

2

TIME-DOMAIN ANALYSIS OF TWO-CONDUCTOR LINES

In this chapter we study the time-domain solution of transmission lines composed of two conductors of length $\mathcal{L}$ as shown in Fig. 2.1. The two conductors are assumed to be parallel to the z axis of a rectangular coordinate system and of uniform cross section along their lengths. The source driving the line, $V_S(t)$, may have some general waveshape in time, although we generally concentrate on periodic waveforms of trapezoidal shape, as in Fig. 1.1, representing digital clock and data signals. The source will be represented as a Thevenin equivalent circuit consisting of a voltage source $V_S(t)$ in series with a source resistor represented by R_S. The load will be represented by a resistance R_L, although PSPICE can be used to investigate lines terminated at either end in nonlinear, dynamic terminations. The voltage and current along the line will be a function of position along the line, z, and time, t, $V(z, t)$ and $I(z, t)$. Given the parameters of the source waveform and the values of the source and load terminations along with the dimensions and properties of the line, we will be interested in determining the solutions for the voltage and current waveforms at the input and output of the line, $V(0, t)$, $V(\mathcal{L}, t)$, $I(0, t)$, and $I(\mathcal{L}, t)$.

Transmission Lines in Digital and Analog Electronic Systems: Signal Integrity and Crosstalk, By Clayton R. Paul
Copyright © 2010 John Wiley & Sons, Inc.

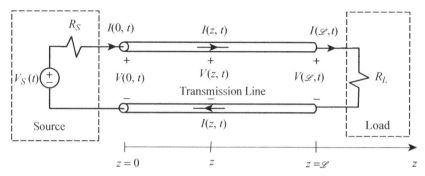

FIGURE 2.1. General configuration of the two-conductor transmission lines to be studied.

2.1   THE TRANSVERSE ELECTROMAGNETIC (TEM) MODE OF PROPAGATION AND THE TRANSMISSION-LINE EQUATIONS

We consider primarily *lossless lines* in which the two conductors and the surrounding medium are considered to be lossless. In Chapter 3 we investigate briefly the effects of losses in the conductors and the surrounding medium. The two conductors are parallel to each other and the z axis and have general cross sections that are uniform along their z-length axis, as shown in Fig. 2.2. They are said to be *uniform lines*.

If we apply a voltage between the two conductors, an electric field, $\vec{E}_T$, will be developed between them that lies solely in the x–y plane. If we pass a current down one conductor and return on the other conductor, a magnetic field, $\vec{H}_T$, will be developed (by the right-hand rule) that lies solely in the x–y plane and passes through the loop between them. This is said to be the *transverse electromagnetic* (TEM) *mode of propagation* that propagates down the line from the source to the load. This electromagnetic field propagates the source signal, $V_S(t)$, from the source to the load. These are said to be *plane waves* since the electromagnetic fields lie in the x–y plane *transverse* to the direction of propagation.

The voltage places a plus charge on the upper conductor and an equal but negative charge on the lower conductor and hence separates charge. Hence the two conductors separate charge and therefore represent a capacitance. We assume that the two conductors are infinitely long so that we can neglect fringing of the fields at the two ends of the conductors. Similarly, the current passing down the upper conductor and returning on the lower conductor causes a magnetic field to penetrate the loop between the two conductors and hence represents an inductance. Since the two conductors are assumed to be infinitely long, we can represent them with a *per-unit-length* capacitance $c(\mathrm{F/m})$ and a *per-unit-length* inductance $l(\mathrm{H/m})$. Since the line is uniform,

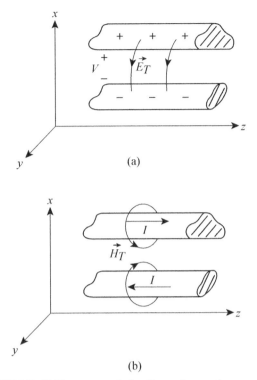

FIGURE 2.2. Uniform transmission lines of general cross section.

we can divide the line into Δz sections that are *electrically small*, as shown in Fig. 2.3, and represent it with a lumped equivalent circuit as shown in Fig. 2.3. Observe that we represent the total inductance and capacitance over each Δz segment with $c\,\Delta z$ and $l\,\Delta z$.

For this lumped equivalent circuit to be valid for all frequencies of the source we, let $\Delta z \to 0$ and obtain the *transmission-line equations* from Fig. 2.4. First, write Kirchhoff's voltage law (KVL) as

$$V(z+\Delta z, t) - V(z,t) = -l\,\Delta z\frac{\partial I(z,t)}{\partial t} \qquad (2.1a)$$

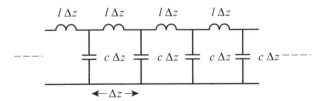

FIGURE 2.3. Representing the line with electrically short segments.

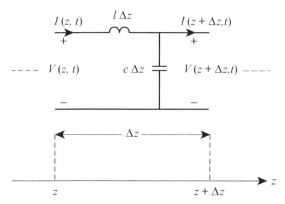

FIGURE 2.4. Per-unit-unit-length equivalent circuit.

and write Kirchhoff's current law (KCL) as

$$I(z+\Delta z,t) - I(z,t) = -c\,\Delta z\,\frac{\partial V(z+\Delta z,t)}{\partial t} \qquad (2.1b)$$

Dividing both sides by Δz gives

$$\frac{V(z+\Delta z,t) - V(z,t)}{\Delta z} = -l\frac{\partial I(z,t)}{\partial t} \qquad (2.2a)$$

$$\frac{I(z+\Delta z,t) - I(z,t)}{\Delta z} = -c\frac{\partial V(z+\Delta z,t)}{\partial t} \qquad (2.2b)$$

and letting $\Delta z \to 0$ yields

$$\lim_{\Delta z \to 0}\frac{V(z+\Delta z,t) - V(z,t)}{\Delta z} = \frac{\partial V(z,t)}{\partial z} \qquad (2.3a)$$

$$\lim_{\Delta z \to 0}\frac{I(z+\Delta z,t) - I(z,t)}{\Delta z} = \frac{\partial I(z,t)}{\partial z} \qquad (2.3b)$$

We then obtain the *transmission-line equations*,

$$\boxed{\frac{\partial V(z,t)}{\partial z} = -l\frac{\partial I(z,t)}{\partial t}} \qquad (2.4a)$$

$$\boxed{\frac{\partial I(z,t)}{\partial z} = -c\frac{\partial V(z,t)}{\partial t}} \qquad (2.4b)$$

which are a set of coupled partial differential equations. We can "uncouple" these coupled partial differential equations by differentiating one with respect of z and the other with respect to t:

$$\frac{\partial^2 V(z,t)}{\partial z^2} = -l\frac{\partial^2 I(z,t)}{\partial z \, \partial t} \tag{2.5a}$$

$$\frac{\partial^2 I(z,t)}{\partial t \, \partial z} = -c\frac{\partial^2 V(z,t)}{\partial t^2} \tag{2.5b}$$

and substitute to obtain the uncoupled second-order differential equations which we are to solve:

$$\boxed{\frac{\partial^2 V(z,t)}{\partial z^2} = \underbrace{lc}_{1/v^2}\frac{\partial^2 V(z,t)}{\partial t^2}} \tag{2.6a}$$

$$\boxed{\frac{\partial^2 I(z,t)}{\partial z^2} = \underbrace{cl}_{1/v^2}\frac{\partial^2 I(z,t)}{\partial t^2}} \tag{2.6b}$$

The velocity of propagation of the TEM waves along the line is

$$\boxed{v = \frac{1}{\sqrt{lc}} \quad \text{m/s}} \tag{2.7}$$

so we can obtain one parameter from the other:

$$\boxed{l = \frac{1}{cv^2}} \tag{2.8a}$$

$$\boxed{c = \frac{1}{lv^2}} \tag{2.8b}$$

If the medium surrounding the two conductors is *homogeneous* with uniform permittivity and permeability ε and μ, respectively, we have the identity

$$\boxed{lc = \mu\varepsilon \quad \text{homogeneous medium}} \tag{2.9}$$

in which case

$$v = \frac{1}{\sqrt{\mu\varepsilon}} \qquad \text{homogeneous medium} \tag{2.10}$$

The general solution to the transmission-line equations is

$$V(z, t) = V^+\left(t - \frac{z}{v}\right) + V^-\left(t + \frac{z}{v}\right) \tag{2.11a}$$

$$I(z, t) = \frac{1}{Z_C} V^+\left(t - \frac{z}{v}\right) - \frac{1}{Z_C} V^-\left(t + \frac{z}{v}\right) \tag{2.11a}$$

where Z_C is the *characteristic impedance of the line*:

$$\begin{aligned} Z_C &= \sqrt{\frac{l}{c}} \quad \Omega \\ &= vl \\ &= \frac{1}{vc} \end{aligned} \tag{2.12}$$

The V^+ and V^- are, as yet, undetermined functions but depend on z, t, and v only as $t + z/v$ and $t - z/v$. These functions are determined by the source and load $V_S(t)$, R_S, and R_L. Also note that there is an important negative sign in the solution for the current. The V^+ represents *forward-traveling waves* traveling in the $+z$ direction, whereas the V^- represents *backward-traveling waves* traveling in the $-z$ direction, respectively. So, in general, we have waves of voltage and current (or, equivalently, waves of electric and magnetic fields) traveling back and forth down the line. We see that the voltage and current waves are in general being reflected at the source and at the load, and the combination of these waves determines, the total voltage and current at the source and load ends of the line.

2.2   THE PER-UNIT-LENGTH PARAMETERS

The transmission-line equations are identical in form for all transmission lines. What distinguishes one line cross section from another? All line cross-sectional dimensions and properties are contained in the per-unit-length parameters l and c. So we must first obtain equations for l and c for each specific line cross section. We consider two general types of line cross sections: wire-type lines, consisting of conductors of circular cylindrical cross section (wires), and lines whose conductors have rectangular cross section that are typical on printed circuit boards (PCBs).

2.2.1 Wire-Type Lines

Wire-type lines consist of conductors of circular cylindrical cross section commonly called *wires*. These common configurations are illustrated in Fig. 2.5. Figure 2.5(a) shows a two-wire line, Fig. 2.5(b) shows one wire above an infinite, perfectly conducting ground plane, and Fig. 2.5(c) shows a coaxial cable.

First we determine the dc magnetic field of an infinitely long, isolated wire of radius r_w as shown in Fig. 2.6. The dc magnetic field of an isolated wire both inside and outside the wire can be determined using Ampere's law from Fig. 2.6:

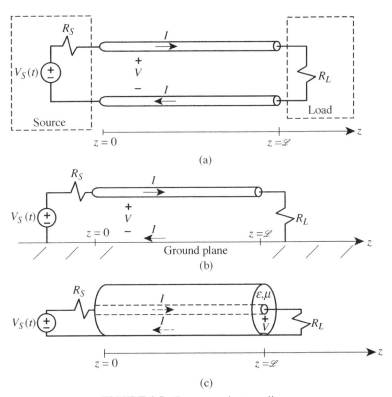

FIGURE 2.5. Common wire-type lines.

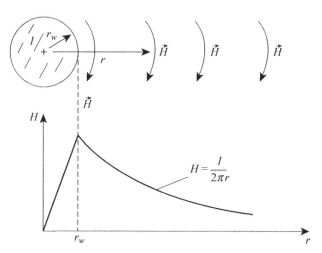

FIGURE 2.6. Dc magnetic field of an isolated wire.

$$\oint_c \vec{\mathbf{H}} \cdot d\vec{\mathbf{l}} = I_{\text{enclosed}} \tag{2.13}$$

Utilizing symmetry, we construct a circular contour c of radius r inside the wire and obtain

$$
\begin{aligned}
\oint_c \vec{\mathbf{H}} \cdot d\vec{\mathbf{l}} &= H_\phi 2\pi r \\
&= I_{\text{enclosed}} \\
&= \frac{I}{\pi r_{\text{w}}^2} \pi r^2 \\
&= I \frac{r^2}{r_{\text{w}}^2}
\end{aligned}
\tag{2.14}
$$

so that

$$
\boxed{
\begin{aligned}
B_\phi &= \mu_0 H_\phi \\
&= \frac{\mu_0 I r}{2\pi r_{\text{w}}^2} \qquad r < r_w
\end{aligned}
}
\tag{2.15}
$$

Constructing a circular contour of radius r outside the wire and utilizing symmetry gives

$$
\begin{aligned}
\oint_c \vec{\mathbf{H}} \cdot d\vec{\mathbf{l}} &= H_\phi 2\pi r \\
&= I_{\text{enclosed}} \\
&= I
\end{aligned}
\tag{2.16}
$$

so that

$$B_\phi = \mu_0 H_\phi$$

$$= \frac{\mu_0 I}{2\pi r} \qquad r > r_w \qquad (2.17)$$

These are plotted in Fig. 2.6.

Next we determine the magnetic flux through a planar surface of length $\mathcal{L}$ that is parallel to the wire and has edges of radii $R_2 > R_1$ from the the wire center axis. This is shown in Fig. 2.7. Gauss's law provides that the magnetic flux through a closed surface is zero; that is, all magnetic field lines must form closed loops:

$$\oint_s \vec{\mathbf{B}} \cdot \vec{d\mathbf{s}} = 0 \qquad (2.18)$$

Form a closed wedge-shaped surface as shown in Fig. 2.7(b). Since the magnetic field is tangent to the ends of this wedge-shaped closed surface and is tangent to S_2, the magnetic flux through the original surface S is equivalent to the flux through the simpler surface S_1:

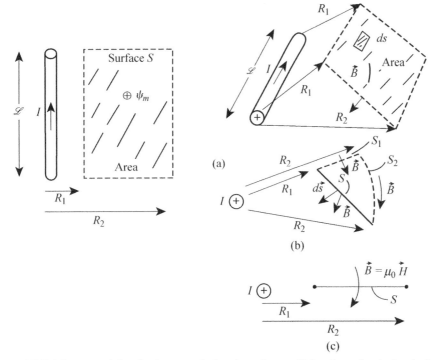

FIGURE 2.7. Determining the dc magnetic flux through a parallel surface of an isolated wire using Gauss's law.

$$\psi = \int_S \vec{\mathbf{B}} \cdot \vec{ds}$$

$$= \int_{S_1} \vec{\mathbf{B}} \cdot \vec{ds} + \underbrace{\int_{S_2} \vec{\mathbf{B}} \cdot \vec{ds}}_{0} + \underbrace{\int_{S_{ends}} \vec{\mathbf{B}} \cdot \vec{ds}}_{0} \tag{2.19}$$

$$= \int_{r=R_1}^{R_2} \frac{\mu_0 I}{2\pi r} dr$$

$$= \frac{\mu_0 I}{2\pi} \ln \frac{R_2}{R_1} \quad \text{Wb/m} \qquad R_2 > R_1$$

This is a fundamental subproblem for determining the per-unit-length inductance of a parallel-wire line.

Another fundamental subproblem is determining the per-unit-length capacitance of a parallel-wire line. First is determining the dc electric field of an infinitely long, isolated wire that carries a dc distribution of charge $q(\text{C/m})$ that is distributed uniformly along its length and around its periphery, as shown in Fig. 2.8(a). Constructing a circular contour of radius r outside the wire as shown in Fig. 2.8(b), assuming that the charge is distributed uniformly along and around the wire, and using Gauss's law gives

$$\oint_S \varepsilon \vec{\mathbf{E}} \cdot \vec{ds} = Q_{\text{enclosed}} \tag{2.20}$$

By symmetry we can determine the electric field at a distance r from the wire axis as

$$\oint_S \varepsilon \vec{\mathbf{E}} \cdot \vec{ds} = \varepsilon E 2\pi r \mathscr{L}$$

$$= Q_{\text{enclosed}} \tag{2.21}$$

$$= q\mathscr{L}$$

thereby giving the electric field directed radially away from the wire as

$$E = \frac{q}{2\pi\varepsilon r} \quad \text{V/m} \quad r > r_{\text{w}}$$

$$E = 0 \qquad\qquad r < r_{\text{w}} \tag{2.22}$$

This is plotted in Fig. 2.8(b).

Next we determine the voltage between two points located at radial distances $R_2 > R_1$ from the axis of the wire as shown in Fig. 2.9. We assume

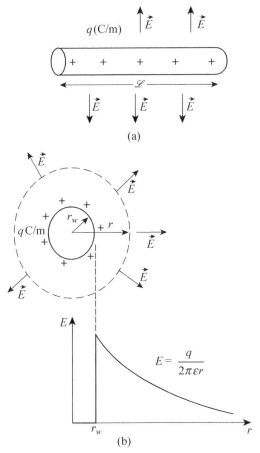

FIGURE 2.8. Determining the dc electric field of an isolated wire.

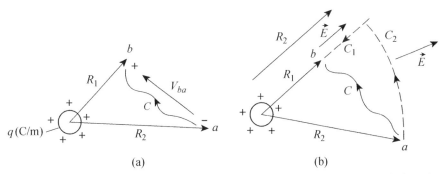

FIGURE 2.9. Determining the voltage between two points that are distances $R_2 > R_1$ from the axis of the wire.

that the point at distance R_1 that is closer to the wire is at the higher potential. The voltage beween the two points is determined as

$$
\begin{aligned}
V_{ba} &= -\int_c \vec{E} \cdot d\vec{l} \\
&= -\int_{c_1} \vec{E} \cdot d\vec{l} - \underbrace{\int_{c_2} \vec{E} \cdot d\vec{l}}_{0} \\
&= -\int_{r=R_2}^{R_1} \frac{q}{2\pi\varepsilon\, r}\, dr \\
&= \frac{q}{2\pi\varepsilon} \ln\frac{R_2}{R_1} \quad \text{V} \qquad R_2 > R_1
\end{aligned}
\tag{2.23}
$$

These two fundamental problems allow us to determine the per-unit-length inductance l and capacitance c of a wire-type line.

First we determine the per-unit-length inductance and capacitance of a two-wire line. Consider two wires of radii r_{w1} and r_{w2} separated by a distance s as shown in Fig. 2.10.

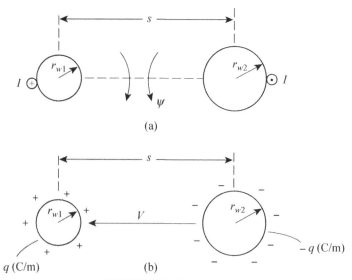

(a)

(b)

FIGURE 2.10. Two-wire line.

Using superposition we obtain the total magnetic flux penetrating the flat surface between the two wires using the two subproblems as shown in Fig. 2.10 (a) as

$$
\begin{aligned}
\psi &= \frac{\mu_0 I}{2\pi} \ln \frac{s - r_{w2}}{r_{w1}} + \frac{\mu_0 I}{2\pi} \ln \frac{s - r_{w1}}{r_{w2}} \\
&= \frac{\mu_0 I}{2\pi} \ln \frac{(s - r_{w1})(s - r_{w2})}{r_{w1} r_{w2}} \\
&\cong \frac{\mu_0 I}{2\pi} \ln \frac{s^2}{r_{w1} r_{w2}} \qquad s \gg r_{w1}, r_{w2}
\end{aligned}
\tag{2.24}
$$

We assume that the wires are of equal radii (as is the practical case) and are sufficiently "widely separated" so that the current and charge are distributed uniformly around the wire peripheries so that the subproblems above are satisfied. Since the wires are assumed to be widely separated so that the subproblems are satisfied, this reduces to

$$
\psi = \frac{\mu_0 I}{\pi} \ln \frac{s}{r_w} \qquad s \gg r_{w1} = r_{w2}
\tag{2.25}
$$

Hence the per-unit-length inductance of the line is

$$
\begin{aligned}
l &= \frac{\psi}{I} \\
&= \frac{\mu_0}{\pi} \ln \frac{s}{r_w} \quad \text{H/m} \qquad s \gg r_w
\end{aligned}
\tag{2.26}
$$

Hence the per-unit-length capacitance of the line can be obtained from

$$
\begin{aligned}
c &= \frac{1}{v_0^2 l} \\
&= \frac{\mu_0 \varepsilon_0}{l} \\
&= \frac{\pi \varepsilon_0}{\ln(s/r_w)} \quad \text{F/m} \quad s \gg r_w
\end{aligned}
\tag{2.27}
$$

If the two wires are "closely spaced," the charges and currents will not be distributed uniformly around the wire peripheries but will concentrate on the facing sides. This is called the *proximity effect*. The exact solution for the per-unit-length inductance is

$$l_{\text{exact}} = \frac{\mu_0}{\pi} \ln \left[\frac{s}{2r_w} + \sqrt{\left(\frac{s}{2r_w} \right)^2 - 1} \right] \qquad \text{H/m} \qquad (2.28)$$

Figure 2.11 shows the ratio of the wide-separation approximation and the exact value. For a ratio of $s/r_w = 4$, the error is only 5.3%. For this ratio one wire would just fit between the two, so this is still "pretty close." For the ratio of $s/r_w = 2$, the wires are just touching.

For the case of one wire above an infinite, perfectly conducting ground plane, the *method of images* can be used to replace the ground plane with an equivalent two-wire problem as shown in Fig. 2.12. All the fields above the ground plane are the same as with the ground plane removed. The magnetic fields penetrating the loop between the wire and the ground plane are one-half those between the two wires and therefore

$$l_{\text{ground plane}} = \frac{1}{2} l_{\text{two wire}}$$

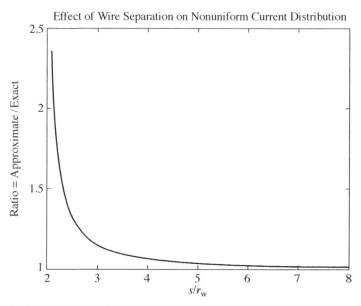

FIGURE 2.11. Comparison of the exact and approximate per-unit-length results for the per-unit-length inductance for a two-wire line.

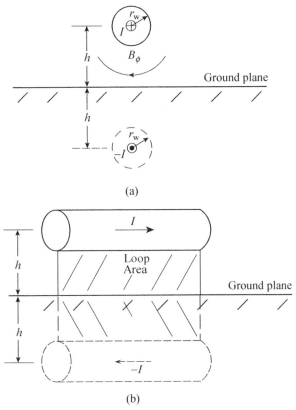

FIGURE 2.12. Transmission line consisting of one wire above an infinite and perfectly conducting ground plane.

Therefore, the per-unit-length inductance for this case is

$$l_{\text{ground plane}} = \frac{\mu_0}{2\pi} \ln\frac{2h}{r_{\text{w}}} \quad \text{H/m} \qquad h \gg r_{\text{w}} \qquad (2.29)$$

This problem can be viewed as being two capacitances in series. Since capacitance in series adds like resistors in parallel, we have

$$c_{\text{ground plane}} = 2c_{\text{two wires}}$$
$$= \frac{2\pi\varepsilon_0}{\ln(2h/r_{\text{w}})} \quad \text{F/m} \qquad h \gg r_{\text{w}} \qquad (2.30)$$

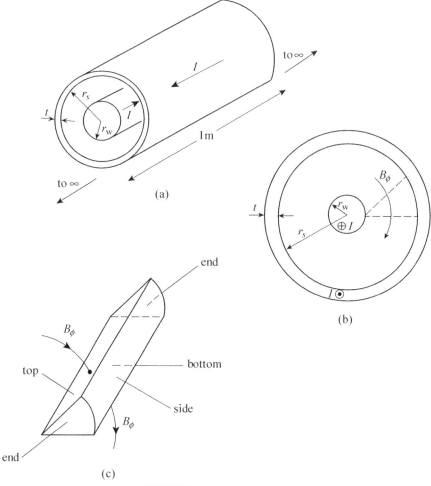

FIGURE 2.13. Coaxial cable.

The final type of wire line is the coaxial cable shown in Fig. 2.13. The coaxial cable consists of a wire of radius r_w centered on the axis of an overall shield of interior radius r_s and thickness t. To determine the per-unit-length inductance l of the cable we must determine, using the preceding subproblem, the magnetic flux penetrating the flat surface between the surface of the interior wire and the interior surface of the shield. We have two choices, as shown in Fig. 2.13(b). The simplest is the direct surface that is perpendicular to the interior wire and the interior of the shield. By symmetry, the magnetic flux is perpendicular to that surface and we obtain

$$\psi = \int_S \vec{\mathbf{B}} \cdot d\vec{s}$$

$$= \int_{z=0}^{1m} \int_{r=r_w}^{r=r_s} B_\phi \underbrace{drdz}_{ds}$$

$$= \int_{z=0}^{1m} \int_{r=r_w}^{r=r_s} \frac{\mu_0 I}{2\pi r} \underbrace{drdz}_{ds}$$

$$= \frac{\mu_0 I}{2\pi} \ln \frac{r_s}{r_w}$$

(2.31)

There is another choice for this surface, but does this give a different answer? Constructing a wedge-shaped surface and utilizing Gauss's law:

$$\oint_S \vec{\mathbf{B}} \cdot d\vec{s} = 0$$

yields

$$\oint_S \vec{\mathbf{B}} \cdot d\vec{s} = \int_{top} \vec{\mathbf{B}} \cdot d\vec{s} + \int_{bottom} \vec{\mathbf{B}} \cdot d\vec{s}$$

$$+ \underbrace{\int_{side} \vec{\mathbf{B}} \cdot d\vec{s}}_{0} + \underbrace{\int_{left\ end} \vec{\mathbf{B}} \cdot d\vec{s}}_{0} + \underbrace{\int_{right\ end} \vec{\mathbf{B}} \cdot d\vec{s}}_{0} = 0$$

Hence we see that

$$\int_{top} \vec{\mathbf{B}} \cdot d\vec{s} = -\int_{bottom} \vec{\mathbf{B}} \cdot d\vec{s}$$

Therefore, the per-unit-length inductance is

$$l = \frac{\psi}{I}$$

$$= \frac{\mu_0}{2\pi} \ln \frac{r_s}{r_w} \qquad H/m$$

(2.32)

The per-unit-length capacitance can be found from this as

$$c = \frac{\mu_0 \varepsilon_0 \varepsilon_r}{l}$$

$$= \frac{2\pi \varepsilon_0 \varepsilon_r}{\ln(r_s/r_w)} \qquad F/m$$

(2.33)

where the coaxial cable is cable is filled with a dielectric having a relative permittivity ε_r. Observe that this result is exact and is not affected by the proximity effect.

2.2.2 Lines of Rectangular Cross Section

The per-unit-length parameters of conductors of rectangular cross section (PCB lands) are very difficult to derive. These results are generally obtained by numerical methods and are given in terms of the characteristic impedance of the line and the effective relative permittivity of the line, ε_r'.

The first case is that of a stripline shown in Fig. 2.14. This is typically found on innerplane PCBs where lands are buried between ground and power planes. The center strip (land) is of width w and is midway between two planes that are separated a distance s. The space between the planes and surrounding the strip is filled with a homogeneous dielectric having a relative permittivity ε_r. Assuming a zero thickness strip, $t = 0$, the per-unit-length inductance is

$$l = \frac{30\pi}{v_0} \frac{1}{(w_e/s) + 0.441} \qquad \text{H/m} \qquad (2.34)$$

and the effective width of the center conductor is

$$\frac{w_e}{s} = \begin{cases} \dfrac{w}{s} & \dfrac{w}{s} \geq 0.35 \\ \dfrac{w}{s} - \left(0.35 - \dfrac{w}{s}\right)^2 & \dfrac{w}{s} \leq 0.35 \end{cases} \qquad (2.35)$$

The per-unit-length capacitance can be found from the inductance since the surrounding medium is homogeneous:

$$\begin{aligned} c &= \frac{\varepsilon_r}{lv_0^2} \\ &= \frac{\varepsilon_r}{30\pi v_0} \left(\frac{w_e}{s} + 0.441\right) \qquad \text{F/m} \end{aligned} \qquad (2.36)$$

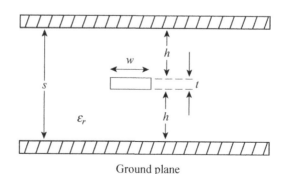

Ground plane

FIGURE 2.14. Stripline.

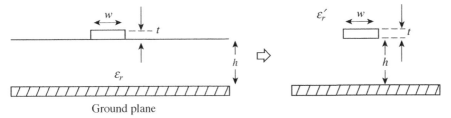

FIGURE 2.15. Microstripline.

The second configuration is that of a microstrip where a land of width w is situated on top of a substrate that has a relative permittivity of ε_r and a thickness h, as shown in Fig. 2.15. This is typical of inner plane PCBs, where lands are found on the top surface of a PCB that has inner planes buried in it. The substrate is on top of an infinite, perfectly conducting ground plane. Assuming a zero thickness strip, $t = 0$, the per-unit-length inductance is

$$
l = \begin{cases}
\dfrac{60}{v_0}\ln\left(\dfrac{8h}{w} + \dfrac{w}{4h}\right) & \text{H/m} & \dfrac{w}{h} \leq 1 \\[3mm]
\dfrac{120\pi}{v_0}\left[\dfrac{w}{h} + 1.393 + 0.667\ln\left(\dfrac{w}{h} + 1.444\right)\right]^{-1} & \text{H/m} & \dfrac{w}{h} \geq 1
\end{cases}
$$

(2.37)

The effective relative permittivity is

$$
\varepsilon_r' = \frac{\varepsilon_r + 1}{2} + \frac{\varepsilon_r - 1}{2}\frac{1}{\sqrt{1 + 12(h/w)}}
$$

(2.38)

This effective relative permittivity accounts for the fact that the electric field lines are partly in air and partly in the substrate dielectric. If this inhomogeneous medium (air and the dielectric) is replaced with a homogeneous medium having an effective relative permittivity of ε_r' as shown, all properties of the line remain unchanged. But the homogeneous medium problem is much easier to analyze. Hence the per-unit-length capacitance is

$$
c = \frac{\varepsilon_r'}{l v_0^2}
$$

$$
= \begin{cases}
\dfrac{\varepsilon_r'}{60 v_0 \ln(8h/w + w/4h)} & \text{F/m} & \dfrac{w}{h} \leq 1 \\[3mm]
\dfrac{\varepsilon_r'}{120\pi v_0}\left[\dfrac{w}{h} + 1.393 + 0.667\ln\left(\dfrac{w}{h} + 1.444\right)\right] & \text{F/m} & \dfrac{w}{h} \geq 1
\end{cases}
$$

(2.39)

The final configuration is that found on low-cost PCBs consisting of two lands on top of a substrate having a relative permittivity of ε_r. The lands are of

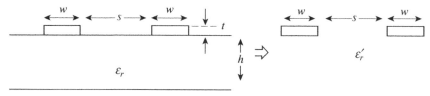

FIGURE 2.16. PCB.

width w and assumed thickness $t = 0$ and are separated by an edge-to-edge separation s as shown in Fig. 2.16.

$$
l = \begin{cases}
\dfrac{120}{v_0} \ln\left(2\dfrac{1+\sqrt{k}}{1-\sqrt{k}} \right) & \text{H/m} \quad \dfrac{1}{\sqrt{2}} \le k \le 1 \\[4mm]
\dfrac{377\pi}{v_0 \ln\left(2\dfrac{1+\sqrt{k'}}{1-\sqrt{k'}} \right)} & \text{H/m} \quad 0 \le k \le \dfrac{1}{\sqrt{2}}
\end{cases}
\tag{2.40a}
$$

where k is

$$
k = \frac{s}{s+2w}
\tag{2.40b}
$$

and $k' = \sqrt{1-k^2}$. The effective relative permittivity is

$$
\varepsilon_r' = \frac{\varepsilon_r+1}{2} \left\{ \tanh\left(0.775\ln\frac{h}{w} + 1.75\right) \right.
$$
$$
\left. + \frac{kw}{h}[0.04 - 0.7k + 0.01(1 - 0.1\varepsilon_r)(0.25 + k)] \right\}
\tag{2.41}
$$

which again accounts for the fact that the electric field lines are partly in air and partly in the substrate dielectric. If this inhomogeneous medium (air and the dielectric) is replaced with a homogeneous medium having an effective relative permittivity of ε_r', all properties of the line remain unchanged. Hence the per-unit-length capacitance is

$$
c = \frac{\varepsilon_r'}{l v_0^2}
$$

$$
= \begin{cases}
\dfrac{\varepsilon_r'}{120\, v_0 \ln\left(2\dfrac{1+\sqrt{k}}{1-\sqrt{k}} \right)} & \text{F/m} \quad \dfrac{1}{\sqrt{2}} \le k \le 1 \\[6mm]
\dfrac{\varepsilon_r' \ln\left(2\dfrac{1+\sqrt{k'}}{1-\sqrt{k'}} \right)}{377\pi\, v_0} & \text{F/m} \quad 0 \le k \le \dfrac{1}{\sqrt{2}}
\end{cases}
\tag{2.42}
$$

2.3 THE GENERAL SOLUTIONS FOR THE LINE VOLTAGE AND CURRENT

The general solutions for the transmission-line voltage and current are given in (2.11) in terms of traveling waves. First look at the load, $z = \mathcal{L}$. The one-way time delay is

$$T_D = \frac{\mathcal{L}}{v} \tag{2.43}$$

At the load we have $z = \mathcal{L}$:

$$V(\mathcal{L}, t) = V^+(t - T_D) + V^-(t + T_D)$$
$$I(\mathcal{L}, t) = \frac{1}{Z_C} V^+(t - T_D) - \frac{1}{Z_C} V^-(t + T_D) \tag{2.44}$$

For a resistive load of R_L, Ohm's law relates the *total* load voltage and current as

$$\frac{V(\mathcal{L}, t)}{I(\mathcal{L}, t)} = R_L \tag{2.45}$$

Suppose that we *only* have a forward-traveling (incoming) wave at the load:

$$\frac{V(\mathcal{L}, t)}{I(\mathcal{L}, t)} = \frac{V^+(t - T_D)}{(1/Z_C)V^+(t - T_D)} = Z_C \tag{2.46}$$

If the load is *matched* (i.e., $R_L = Z_C$), we will only have a forward-traveling (incoming) wave at the load and there will be no reflected waves at the load if $R_L = Z_C$. But for some general load that is not matched, $R_L \neq Z_C$, we must have an incident (forward-traveling) wave and a reflected (backward-traveling) wave at the load in order to satisfy Ohm's law.

Define the *voltage reflection coefficient at the load* as the ratio of the reflected and incident voltage waves:

$$\Gamma_L = \frac{V^-(t + T_D)}{V^+(t - T_D)} \tag{2.47}$$

If we know the load reflection coefficient, Γ_L, we can determine the reflected voltage wave knowing the incident voltage wave. The *total* voltage and current at the load can then be written in terms of the load reflection coefficient as

$$\begin{array}{l} V(\mathscr{L}, t) = V^+(t - T_D)(1 + \Gamma_L) \\ I(\mathscr{L}, t) = \dfrac{1}{Z_C} V^+(t - T_D)(1 - \Gamma_L) \end{array} \tag{2.48}$$

Taking the ratio of these two relations gives

$$\begin{aligned} \frac{V(\mathscr{L}, t)}{I(\mathscr{L}, t)} &= R_L \\ &= Z_C \frac{1 + \Gamma_L}{1 - \Gamma_L} \end{aligned} \tag{2.49}$$

Solving this gives the *voltage reflection coefficient at the load* as

$$\Gamma_L = \frac{R_L - Z_C}{R_L + Z_C} \tag{2.50}$$

Observe that since there is a minus sign in the current relation, the current reflection coefficient is the negative of the voltage reflection coefficient:

$$\Gamma_L|_{\text{current}} = -\Gamma_L|_{\text{voltage}} \tag{2.51}$$

The process of reflection at the load is like a mirror: the reflected wave is coming out of the mirror and the incident wave is going in, as illustrated in Fig. 2.17. *The total voltage is the sum of the incident and reflected waves.*

Now we investigate what happens at the source. The voltage or current wave that was reflected at the load travels back to the source in another time interval of T_D, where it is reflected with a voltage reflection coefficient of

$$\Gamma_S = \frac{R_S - Z_C}{R_S + Z_C} \tag{2.52}$$

and sent back to the load. The current reflection coefficient at the source is, again, the negative of the voltage reflection coefficient at the source:

$$\Gamma_S|_{\text{current}} = -\Gamma_S|_{\text{voltage}} \tag{2.53}$$

Finally, we obtain the wave sent out initially. We reason that when the source voltage is turned on initially, an initial forward-traveling wave is sent

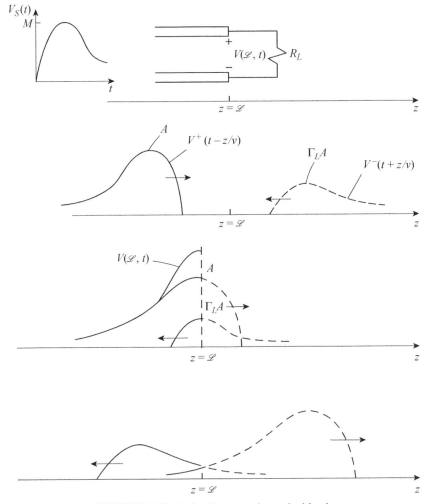

FIGURE 2.17. Reflection at a mismatched load.

out toward the load. This initial wave will take a time delay of T_D to get to the load. Any reflections of this initial wave at a mismatched load will require another one-way time delay of T_D to get back to the source. Hence no reflected wave will have arrived at the source over the time interval $0 < t < 2T_D$. So the total voltage at the source is just the forward-traveling wave sent out initially, and hence the ratio of the total voltage to total current at the source end of the line, $z = 0$, will just be

$$\frac{V(0,t)}{I(0,t)} = \frac{V^+(t-0)}{(1/Z_C)V^+(t-0)} = Z_C \qquad 0 < t < 2T_D \qquad (2.54)$$

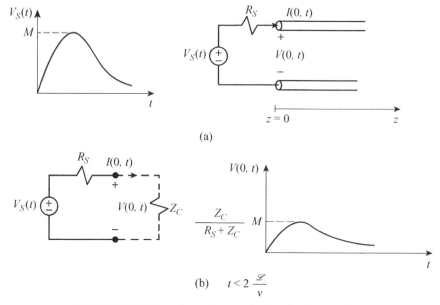

FIGURE 2.18. Input impedance to the line for $0 \le t < 2T_D$.

So the input impedance to the line appears to be Z_C but *only for* $t \le 2T_D$. Hence we can calculate the initially sent out voltage and current waves from

$$V_{\text{init}} = \frac{Z_C}{R_S + Z_C} V_S(t) \qquad (2.55)$$

$$I_{\text{init}} = \frac{V_S(t)}{R_S + Z_C} \qquad (2.56)$$

as illustrated in Fig. 2.18.

2.4   WAVE TRACING AND REFLECTION COEFFICIENTS

So we trace the waves, noting that the total voltage or current at any point on the line and at any time is the sum of the forward- and backward-traveling waves *at that point and at that time* on the line.

EXAMPLE

Sketch the line voltage for the problem of Fig. 2.19 at various positions on the line for fixed times.

First we perform initial computations for the voltage:

$$T_D = \frac{\mathcal{L}}{v}$$

$$= 2\mu s$$

$$V_{init} = \frac{Z_C}{R_S + Z_C} V_S(t)$$

$$= \left(\frac{50}{0 + 50}\right)(30)$$

$$= 30V$$

$$\Gamma_S = \frac{R_S - Z_C}{R_S + Z_C} \qquad \Gamma_L = \frac{R_L - Z_C}{R_L + Z_C}$$

$$= \frac{0 - 50}{0 + 50} \qquad = \frac{100 - 50}{100 + 50}$$

$$= -1 \qquad = \frac{1}{3}$$

Next, sketch the waves on the line at various times, as shown in Fig. 2.20 and then sketch the load voltage $V(\mathcal{L}, t)$ versus time as shown in Fig. 2.21.

Next, we sketch the input current to the line versus time $I(0, t)$ as shown in Fig. 2.22. First, perform the initial computations for the current:

$$I_{init} = \frac{V_S}{R_S + Z_C}$$

$$= \frac{30}{0 + 50}$$

$$= 0.6 \text{ A}$$

$$\Gamma_S|_{current} = -\Gamma_S|_{voltage}$$

$$= +1$$

$$\Gamma_L|_{current} = -\Gamma_L|_{voltage}$$

$$= -\frac{1}{3}$$

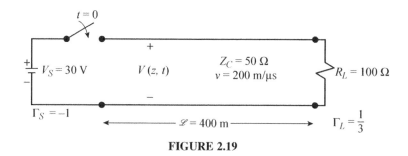

FIGURE 2.19

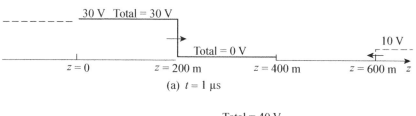

(a) $t = 1$ μs

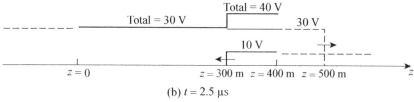

(b) $t = 2.5$ μs

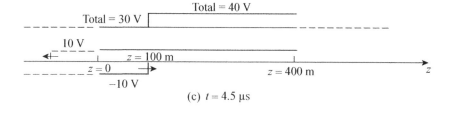

(c) $t = 4.5$ μs

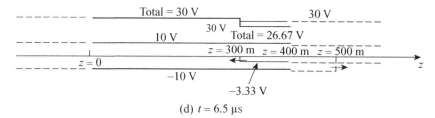

(d) $t = 6.5$ μs

FIGURE 2.20

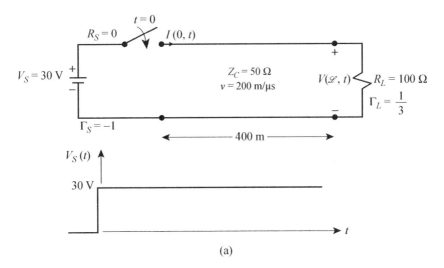

(a)

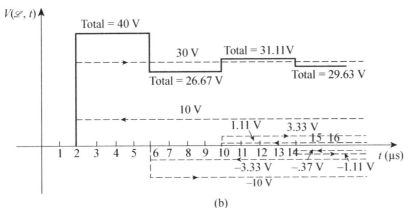

(b)

FIGURE 2.21

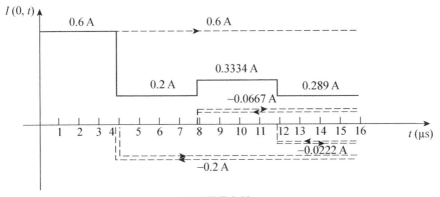

FIGURE 2.22

Always verify that your solution is approaching the steady-state values, which for this example are $V(\mathcal{L}, t)_{\text{steady state}} = 30$ V and $I(0, t)_{\text{steady state}} = 0.3$ A.

EXAMPLE

Sketch the voltage at the input and the output of the line versus time for the problem of Fig. 2.23. This problem illustrates the case where the source voltage waveform, $V_S(t)$, is a pulse of finite duration that is equal to the one-way time delay of the line, $T_D = \frac{\mathcal{L}}{v}$, and hence the incident and reflected pulses at a termination overlap in time and combine. However, since the pulse durations are equal to the one-way duration, they are short enough in time that they do not overlap with and combine with other reflections from those of the reflections from other line endpoints.

First perform the initial computations for the voltages:

$$T_D = \frac{\mathcal{L}}{v}$$

$$= 1 \text{ ns}$$

$$V_{\text{init}} = \frac{Z_C}{R_S + Z_C} V_S(t)$$

$$= \left(\frac{100}{300 + 100} \right)(20)$$

$$= 5 \text{ V}$$

$$\Gamma_S = \frac{R_S - Z_C}{R_S + Z_C}$$

$$= \frac{300 - 100}{300 + 100}$$

$$= \frac{1}{2}$$

$$\Gamma_L = \frac{R_L - Z_C}{R_L + Z_C}$$

$$= \frac{\infty - 100}{\infty + 100}$$

$$= +1$$

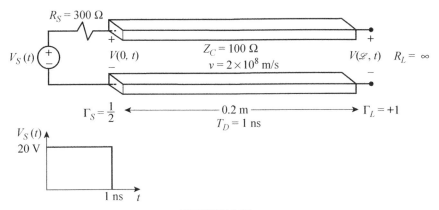

FIGURE 2.23

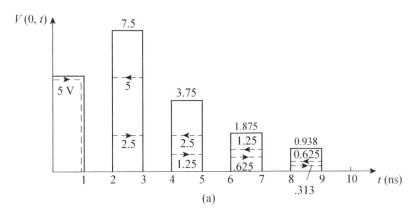

(a)

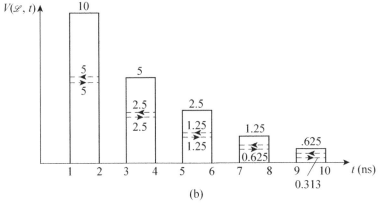

(b)

FIGURE 2.24

Next sketch the input voltage $V(0, t)$ and the load voltage $V(\mathscr{L}, t)$ as shown in Fig. 2.24. Verify that the solutions are approaching the steady-state values, which for this case are $V(0, t)_{\text{steady state}} = 0$ and $V(\mathscr{L}, t)_{\text{steady state}} = 0$.

EXAMPLE

Sketch the voltage at the input, $V(0, t)$, and the current at the output, $I(\mathcal{L}, t)$, of the line versus time for the problem of Fig. 2.25. This problem illustrates the case where the source voltage waveform, $V_S(t)$, is a pulse of finite duration that is several one-way time delays of the line, $T_D = \frac{\mathcal{L}}{v}$, in duration. Hence the incident and reflected pulses from opposite terminations overlap in time and combine to give very complicated total waveshapes at those terminations.

First perform the initial computations:

$$Z_C = \sqrt{\frac{l}{c}}$$

$$= \sqrt{\frac{0.25 \times 10^{-6}}{100 \times 10^{-12}}}$$

$$= 50\,\Omega$$

$$v = \frac{1}{\sqrt{lc}}$$

$$= \frac{1}{\sqrt{(0.25 \times 10^{-6})(100 \times 10^{-12})}}$$

$$= 200\,\text{m}/\mu\text{s}$$

$$T_D = \frac{\mathcal{L}}{v}$$

$$= 2\,\mu\text{s}$$

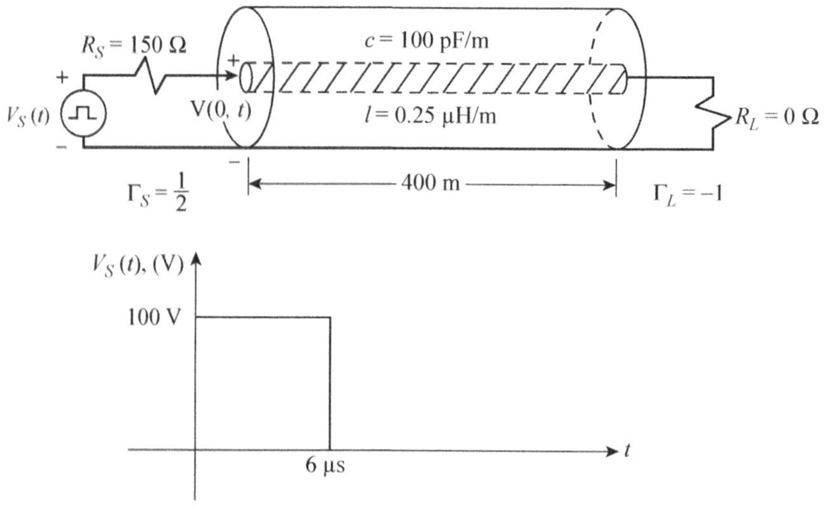

FIGURE 2.25

Perform the initial computations for the voltage:

$$V_{\text{init}} = \frac{Z_C}{R_S + Z_C} V_S(t)$$

$$= \left(\frac{50}{150 + 50}\right)(100)$$

$$= 25 \text{ V}$$

$$\Gamma_S = \frac{R_S - Z_C}{R_S + Z_C} \qquad \Gamma_L = \frac{R_L - Z_C}{R_L + Z_C}$$

$$= \frac{150 - 50}{150 + 50} \qquad = \frac{0 - 50}{0 + 50}$$

$$= \frac{1}{2} \qquad = -1$$

Sketch the input voltage to the line, $V(0, t)$, as shown in Fig. 2.26. Verify that the solution is approaching the steady-state value, which for this case is $V(0, t)_{\text{steady state}} = 0$.

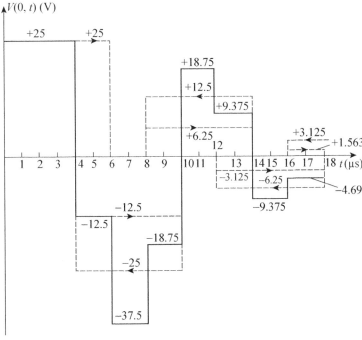

FIGURE 2.26

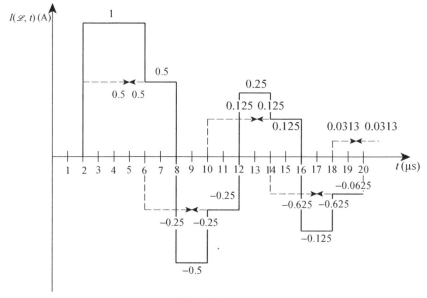

FIGURE 2.27

Next perform the initial computations for the current:

$$I_{\text{init}} = \frac{1}{R_S + Z_C} V_S(t)$$

$$= \left(\frac{1}{150 + 50}\right)(100)$$

$$= 0.5 \text{ A}$$

$$\Gamma_S|_{\text{current}} = -\Gamma_S|_{\text{voltage}} \qquad \Gamma_L|_{\text{current}} = -\Gamma_L|_{\text{voltage}}$$

$$= -\frac{1}{2} \qquad\qquad\qquad = +1$$

Sketch the load current for the line, $I(\mathscr{L}, t)$, as shown if Fig. 2.27. Verify that the solution is approaching the steady-state value, which for this case is $I(\mathscr{L}, t)_{\text{steady state}} = 0$.

We can follow through the wave tracing above to obtain a series solution for the terminal voltages as (you should verify these)

$$V(0, t) = \frac{Z_C}{R_S + Z_C} \left[V_S(t) + (1 + \Gamma_S)\Gamma_L V_S(t - 2T_D) \right.$$

$$+ (1 + \Gamma_S)(\Gamma_S \Gamma_L)\Gamma_L V_S(t - 4T_D) \qquad (2.57a)$$

$$\left. + (1 + \Gamma_S)(\Gamma_S \Gamma_L)^2 \Gamma_L V_S(t - 6T_D) + \cdots \right]$$

and

$$V(\mathcal{L}, t) = \frac{Z_C}{R_S + Z_C} (1 + \Gamma_L) \left[V_S(t - T_D) + (\Gamma_S \Gamma_L) V_S(t - 3T_D) \right.$$

$$+ (\Gamma_S \Gamma_L)^2 V_S(t - 5T_D)$$

$$\left. + (\Gamma_S \Gamma_L)^3 V_S(t - 7T_D) + \cdots \right]$$

$$(2.57b)$$

Observe in these expressions that the total voltages at the input and output to the transmission line are combinations of the source waveform, $V_S(t)$, that are *delayed by two time delays*. Also note that the magnitudes of the source and load reflection coefficients are less than or equal to unity:

$$|\Gamma_S| \leq 1 \qquad (2.58a)$$

and

$$|\Gamma_L| \leq 1 \qquad (2.58b)$$

Also observe that the total load voltage, $V(\mathcal{L}, t)$, in (2.57b) is a sum of delayed replicas of $V_S(t)$ multiplied by products of the source and load reflection coefficients, $(\Gamma_S \Gamma_L)^n$, which are also progressively less than unity. Hence Table 2.1 for $V_S(t) = Mu(t)$ can be constructed. If the source resistor is less than the characteristic impedance, $R_S < Z_C$, and the load resistor is greater than the load resistor, $R_L > Z_C$, or vice versa, the source and load reflection coefficients are of *opposite sign*. Hence the the resulting load voltage will have a portion added to it and subtracted from it, resulting in *oscillations*. On the other hand, if the source and load resistors are both less than the characteristic impedance, $R_S < Z_C$, $R_L < Z_C$, or are both greater than the characteristic impedance, $R_S > Z_C$, $R_L > Z_C$, the source and load reflection coefficients are

TABLE 2.1. Effects of the Signs of the Reflection Coefficients on the Load Voltage

Γ_S	Γ_L	Load Voltage Waveform
$\overset{-}{R_S} < Z_C$ $\overset{+}{R_S} > Z_C$	$\overset{+}{R_L} > Z_C$ $\overset{-}{R_L} < Z_C$	
$\overset{+}{R_S} > Z_C$ $\overset{-}{R_S} < Z_C$	$\overset{+}{R_L} > Z_C$ $\overset{-}{R_L} < Z_C$	

of the *same sign* and the load voltage voltage will steadily build up to its steady-state value.

This explains why a typical digital transmission system, where a low-impedance CMOS driver drives a transmission line that is terminated in a high-input impedance CMOS receiver, will have oscillations (called *overshoot* and *undershoot*) that can result in logic errors.

2.5   THE SPICE (PSPICE) Exact Transmission-Line Model

SPICE (PSPICE) contains an *exact model* for a *two-conductor lossless transmission line*. The advantage of this model is that (1) it is simple and fast to use, and (2) you can use *any terminations*, inductors, capacitors, or nonlinear resistors such as diodes and transistors and see the exact solution for the voltages and currents vs. time at the ends of the line!

SPICE (PSPICE) implements the exact transmission-line solution in a discrete, "bootstrapping" manner. Consider the exact solution of the transmission-line equations:

$$V(z,t) = V^+\left(t - \frac{z}{v}\right) + V^-\left(t + \frac{z}{v}\right)$$

$$Z_C I(z,t) = V^+\left(t - \frac{z}{v}\right) - V^-\left(t + \frac{z}{v}\right)$$

(2.59)

Evaluate them at the source end, $z = 0$:

$$V(0,t) = V^+(t) + V^-(t)$$
$$Z_C I(0,t) = V^+(t) - V^-(t)$$

(2.60)

Evaluate them at the load end, $z = \mathcal{L}$:

$$V(\mathcal{L},t) = V^+(t - T_D) + V^-(t + T_D)$$
$$Z_C I(\mathcal{L},t) = V^+(t - T_D) - V^-(t + T_D)$$

(2.61)

Add and subtract them:

$$V(0,t) + Z_C I(0,t) = 2V^+(t)$$
$$V(0,t) - Z_C I(0,t) = 2V^-(t)$$

(2.62a)

$$V(\mathcal{L},t) + Z_C I(\mathcal{L},t) = 2V^+(t - T_D)$$
$$V(\mathcal{L},t) - Z_C I(\mathcal{L},t) = 2V^-(t + T_D)$$

(2.62b)

Time shift and rearrange:

$$V(0,t) = Z_C I(0,t) + 2V^-(t)$$
$$V(\mathcal{L},t) = -Z_C I(\mathcal{L},t) + 2V^+(t - T_D)$$

(2.63a)

$$V(0,t - T_D) + Z_C I(0,t - T_D) = 2V^+(t - T_D)$$
$$V(\mathcal{L},t - T_D) - Z_C I(\mathcal{L},t - T_D) = 2V^-(t)$$

(2.63b)

This gives

$$V(0,t) = Z_C I(0,t) + \underbrace{V(\mathcal{L},t - T_D) - Z_C I(\mathcal{L},t - T_D)}_{E_0(\mathcal{L},t - T_D)}$$

$$V(\mathcal{L},t) = -Z_C I(\mathcal{L},t) + \underbrace{V(0,t - T_D) + Z_C I(0,t - T_D)}_{E_{\mathcal{L}}(0,t - T_D)}$$

(2.64)

This gives the equivalent circuit shown in Fig. 2.28.

SPICE (PSPICE) uses a bootstrapping or iterative method of solution. On the .TRAN line you can set a *minimum step size*, Δt. Then SPICE solves at $t = 0$ and uses that solution to determine the solution at $t = \Delta t$, and uses that solution to determine the solution at $t = 2\Delta t$, and uses that solution to

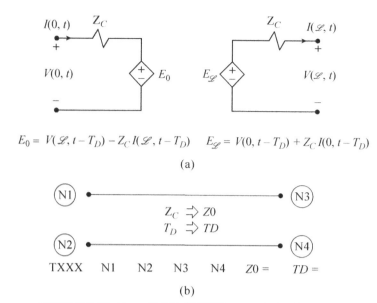

$$E_0 = V(\mathscr{L}, t - T_D) - Z_C I(\mathscr{L}, t - T_D) \qquad E_\mathscr{L} = V(0, t - T_D) + Z_C I(0, t - T_D)$$

(a)

$$TXXX \quad N1 \quad N2 \quad N3 \quad N4 \quad Z0 = \qquad TD =$$

(b)

FIGURE 2.28. Exact SPICE (PSPICE) transmission-line model.

determine the solution at $t = 3\Delta t$, and so on, until it gets to the *final solution time* you specified on the .TRAN line.

Now we solve the previous examples using SPICE (PSPICE).

EXAMPLE

Solve the example in Fig. 2.19 using PSPICE.

See Fig. 2.29. The PSPICE program is

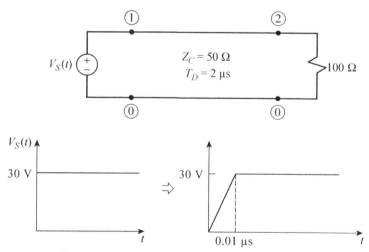

FIGURE 2.29. Example of Fig. 2.19 using PSPICE.

```
EXAMPLE
VS 1 0 PWL(0 0 .01U 30)
T 1 0 2 0 Z0=50 TD=2U
RL 2 0 100
.TRAN    .01U    20U    0    .01U
.PRINT   TRAN    V(2)    I(VS)
.PROBE
.END
```

The results are shown in Fig. 2.30. Check with those obtained by hand.

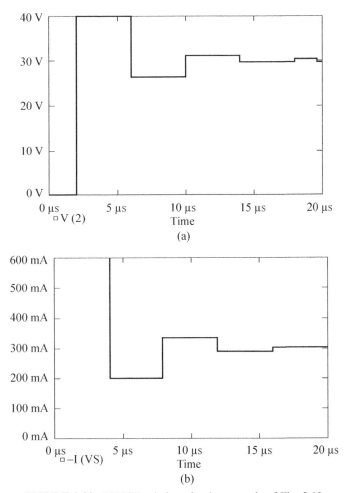

FIGURE 2.30. PSPICE solutions for the example of Fig. 2.19.

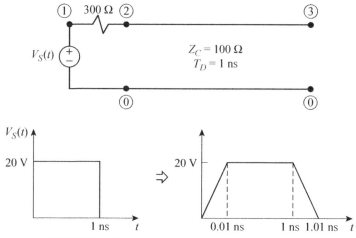

FIGURE 2.31. Example of Fig. 2.23 using PSPICE.

EXAMPLE

Solve the example in Fig. 2.23 using PSPICE.
   See Fig. 2.31. The PSPICE program is

```
EXAMPLE
VS 1 0 PWL(0 0 0.01N 20 1N 20 1.01N 0)
RS 1 2 300
T 2 0 3 0 Z0=100 TD=1N
RL 3 0 1E8
.TRAN 0.01N 10N 0 0.01N
.PRINT TRAN V(2) V(3)
.PROBE
.END
```

The results are shown in Fig. 2.32. Check with those obtained by hand.

EXAMPLE

Solve the example of Fig. 2.25 using PSPICE.
   See Fig. 2.33. The PSPICE program is

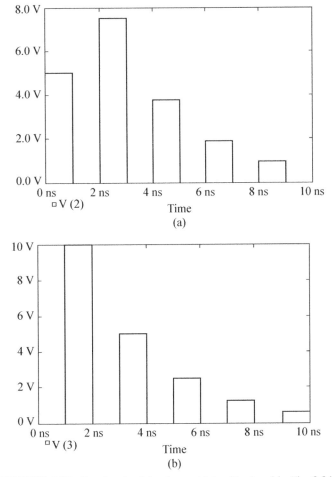

FIGURE 2.32. Results check by those obtained by hand in Fig. 2.24.

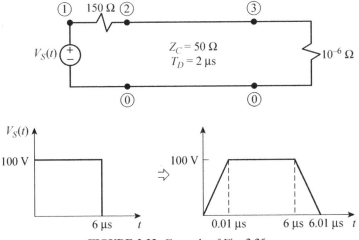

FIGURE 2.33. Example of Fig. 2.25.

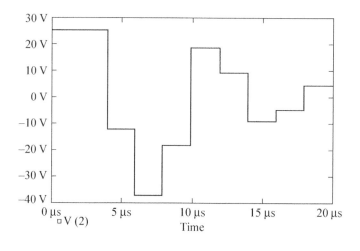

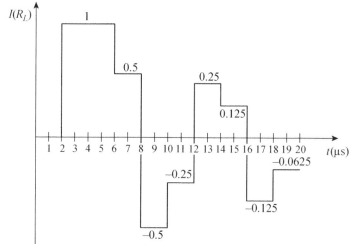

FIGURE 2.34. Results check with those obtained by hand in Figs. 2.26 and 2.27.

```
EXAMPLE
VS 1 0 PWL(0 0 .01U 100 6U 100 6.01U 0)
RS 1 2 150
T 2 0 3 0 Z0=50 TD=2U
RL 3 0 1E-6
.TRAN .01U 20U 0 .01U
.PRINT TRAN V(2) I(RL)
.PROBE
.END
```
The results are shown in Fig. 2.34. Check with those obtained by hand.

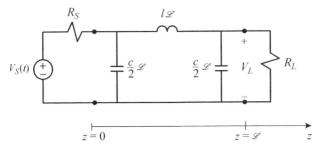

FIGURE 2.35. Lumped-pi equivalent circuit.

2.6 LUMPED-CIRCUIT APPROXIMATE MODELS OF THE LINE

We have seen in Chapter 1 that a lumped equivalent-circuit model of transmission line, such as the lumped Pi circuit shown in Fig. 2.35, can give adequate approximate solutions of the transmission-line equations as long as the line is electrically short at the "significant frequencies" of the source waveform $V_S(t)$ within its bandwidth:

$$\text{BW} = f_{max} \cong \frac{1}{\tau_r} \qquad \tau_r = \tau_f \tag{2.65}$$

In this circuit, the total line inductance, $L = l\mathscr{L}$, is placed in the middle, and the total line capacitance, $C = c\mathscr{L}$, is split and placed on both sides so as to make the model symmetrical, as is the actual line. The criterion for its adequacy is that the total line length must be electrically short at this highest significant frequency:

$$\begin{aligned} \mathscr{L} &< \frac{1}{10}\lambda \\ &= \frac{1}{10}\frac{v}{f_{max}} \end{aligned} \tag{2.66}$$

Substituting gives

$$\boxed{\tau_r > 10T_D} \tag{2.67}$$

But use of this equivalent circuit is unnecessary since an *exact* model of a *lossless* transmission line exists in the PSPICE program.

Since $Z_C = vl$ and $Z_C = 1/vc$, the total inductance and capacitance of the line can be computed in terms of the line characteristic impedance Z_C and the line one-way delay T_D as

$$\boxed{\begin{aligned} L &= l\mathscr{L} \\ &= \frac{Z_C}{v}\mathscr{L} \\ &= Z_C T_D \quad \text{H} \end{aligned}} \tag{2.68a}$$

and

$$
\begin{aligned}
C &= c\mathcal{L} \\
&= \frac{\mathcal{L}}{vZ_C} \\
&= \frac{T_D}{Z_C} \quad F
\end{aligned}
\tag{2.68b}
$$

2.7 EFFECTS OF REACTIVE TERMINATIONS ON TERMINAL WAVEFORMS

So far we have sketched the terminal voltages and currents versus time for resistive terminations. Sketching the corresponding terminal waveforms for reactive and/or nonlinear terminations is very difficult and is unnecessary since we can use the exact model of the transmission line that exists in PSPICE. But now we investigate qualitatively how these dynamic terminations affect these terminal waveforms.

2.7.1 Effect of Capacitive Terminations

Now we investigate the computation of the terminal voltage of the transmission line when the termination is a capacitor as illustrated in Fig. 2.36. Hand sketching the terminal voltages and currents for resistive terminations is easy. But for *reactive terminations* such as capacitive and inductive loads, hand sketching the terminal voltages and currents is more difficult. The Laplace transform of the transmission-line equations makes this easier. Transforming the transmission-line equations *with respect to time t* gives

$$
\begin{aligned}
\frac{dV(z,s)}{dz} &= -slI(z,s) \\
\frac{dI(z,s)}{dz} &= -scV(z,s)
\end{aligned}
\tag{2.69}
$$

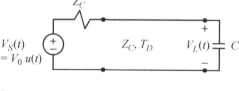

FIGURE 2.36. Termination of a transmission line in a capacitor.

The Laplace transform converts *partial differential equations* into *ordinary differential equations*. The second-order uncoupled equations are

$$\frac{d^2V(z,s)}{dz^2} - s^2 lc V(z,s) = 0$$
$$\frac{d^2I(z,s)}{dz^2} - s^2 lc I(z,s) = 0$$

(2.70)

The solutions are

$$V(z,s) = V^+(s)e^{-s(z/v)} + V^-(s)e^{s(z/v)}$$
$$I(z,s) = \frac{1}{Z_C}V^+(s)e^{-s(z/v)} - \frac{1}{Z_C}V^-(s)e^{s(z/v)}$$

(2.71a)

where $V^+(s)$ and $V^-(s)$ are to be determined by the source waveform and the terminations. Note that the inverse transform is

$$V(z,t) = V^+\left(t - \frac{z}{v}\right) + V^-\left(t + \frac{z}{v}\right)$$
$$I(z,t) = \frac{1}{Z_C}V^+\left(t - \frac{z}{v}\right) - \frac{1}{Z_C}V^-\left(t + \frac{z}{v}\right)$$

(2.71b)

This solution is usually obtained more easily *indirectly*. The Laplace-transformed circuit is shown in Fig. 2.37.

We assumed that the source is matched to make the solution easier, so that $\Gamma_S(s) = 0$. The transformed load reflection coefficient is

$$\Gamma_L(s) = \frac{Z_L(s) - Z_C}{Z_L(s) + Z_C}$$
$$= \frac{(1/sC) - Z_C}{(1/sC) + Z_C}$$

(2.72)

$$= \frac{1 - sT_C}{1 + sT_C}$$

where the time constant is

$$T_C = Z_C C$$

(2.73)

Since the source is matched, there will be no reflections there, so we obtain the load voltage as

$$V_L(t) = (1 + \Gamma_L)\frac{Z_C}{(R_S = Z_C) + Z_C}V_0 u(t - T_D)$$

(2.74)

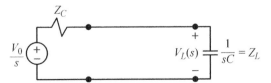

FIGURE 2.37. Laplace-transformed circuit.

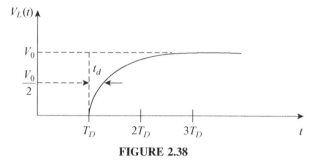

FIGURE 2.38

where $u(t)$ is the unit step function. Transforming this yields

$$V_L(s) = [1 + \Gamma_L(s)]\frac{1}{2}V_0 e^{-sT_D}$$

$$= \frac{1/T_C}{(s + 1/T_C)s}V_0 e^{-sT_D} \tag{2.75}$$

$$= \left[\frac{1}{s} - \frac{1}{(s + 1/T_C)}\right]V_0 e^{-sT_D}$$

The inverse transform of this is

$$V_L(t) = V_0 u(t - T_D) - e^{-(t - T_D)/T_C}V_0 u(t - T_D) \tag{2.76}$$

which is sketched in Fig. 2.38.

This makes sense, of course, because when the unit step voltage arrives at the capacitive load, the capacitor initially looks like a short circuit and then begins charging up to V_0. Notice that in addition to the one-way time delay T_D, there is an additional time delay (to get to half the level) of $V_0 e^{-t/T_C} = \frac{1}{2}V_0$, or

$$t_d = 0.693T_C \tag{2.77}$$

which is due to the time constant

$$T_C = Z_C C \tag{2.78}$$

So to the capacitor the line looks, like a Thévenin equivalent with a source resistance of Z_C.

2.7.2   Effect of Inductive Terminations

Now we investigate the effect of an inductive load in a similar fashion. The Laplace-transformed circuit is shown in Fig. 2.39. Again since the source is assumed matched,

$$\Gamma_S(s) = 0 \tag{2.79}$$

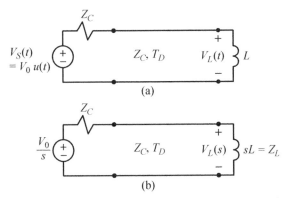

FIGURE 2.39. Transmission line with an inductive load.

The transformed load reflection coefficient is

$$
\begin{aligned}
\Gamma_L(s) &= \frac{Z_L(s) - Z_C}{Z_L(s) + Z_C} \\
&= \frac{sL - Z_C}{sL + Z_C} \\
&= \frac{sT_L - 1}{sT_L + 1}
\end{aligned}
\tag{2.80}
$$

where the time constant is

$$
T_L = \frac{L}{Z_C}
\tag{2.81}
$$

Since the source is matched there will be no reflections there, so we obtain the load voltage as

$$
V_L(t) = (1 + \Gamma_L) \frac{Z_C}{(R_S = Z_C) + Z_C} V_0 u(t - T_D)
\tag{2.82}
$$

where $u(t)$ is the unit step function. Transforming this yields

$$
\begin{aligned}
V_L(s) &= [1 + \Gamma_L(s)] \frac{1}{2} V_0 e^{-sT_D} \\
&= \frac{1}{s + 1/T_L} V_0 e^{-sT_D}
\end{aligned}
\tag{2.83}
$$

The inverse transform of this is

$$
V_L(t) = e^{-(t - T_D)/T_L} V_0 u(t - T_D)
\tag{2.84}
$$

which is sketched in Fig. 2.40.

This makes sense, of course, because when the unit step voltage arrives at the inductive load, the inductor initially looks like an open circuit and then

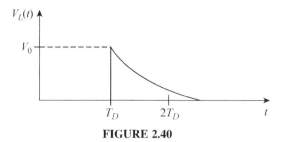

FIGURE 2.40

transitions to a short circuit in the steady state. Since the line is again matched at the source, at the load it again looks to the inductor like a Thévenin equivalent with a source resistance of Z_C.

2.8   MATCHING SCHEMES FOR SIGNAL INTEGRITY

In Section 1.4.1 we saw how mismatches at terminations on the transmission line can cause ringing and other undesirable waveforms that can cause logic errors in digital systems that are significant problems in high-speed digital systems. For example, we saw in Figs. 1.18, 1.19, 1.20, and 1.21 a typical configuration of how a low-impedance CMOS driver attached to a transmission line that is terminated in a high-impedance CMOS receiver can result in overshoot and undershoot termination voltages that can result in causing logic errors in the CMOS receiver.

What causes undershoot and overshoot voltages? To investigate how these come about, we investigate an ideal situation driven by a low-impedance driver terminated in an open-circuit load as shown in Fig. 2.41(a). A hand analysis of this situation is shown in Fig. 2.41(b). The line is driven by an ideal 5-V step function, $V_S(t) = 5u(t)$ V, is shown. The open-circuit load has a voltage reflection coefficient of $\Gamma_L = +1$ and completely reflects the incoming signal, which is also reflected at the source. The series solution for the load voltage can be written as

$$V(\mathcal{L}, t) = \frac{Z_C}{R_S + Z_C}(1 + \Gamma_L)[V_S(t - T_D) + (\Gamma_S\Gamma_L)V_S(t - 3T_D)$$
$$+ (\Gamma_S\Gamma_L)^2 V_S(t - 5T_D) + (\Gamma_S\Gamma_L)^3 V_S(t - 7T_D) + \cdots] \quad (2.85)$$

Note that the *magnitudes* of the reflection coefficients are always less than or equal to unity:

$$|\Gamma_S| \leq 1 \quad \text{and} \quad |\Gamma_L| \leq 1 \quad (2.86)$$

For a low-impedance source, $R_S < Z_C$, Γ_S is negative, and for a high-impedance load, $R_L > Z_C$, Γ_L is positive. So the delayed replica of the source voltage $V_S(t)$ is delayed by multiples of the one-way time delay and multiplied

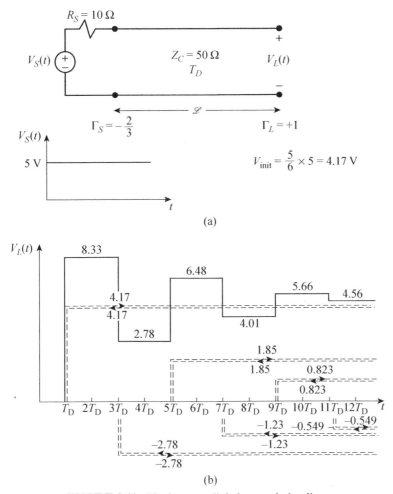

FIGURE 2.41. Ringing on a digital transmission line.

by powers of products of the source and load refelection coefficients, $(\Gamma_S\Gamma_L)^n$, which alternate in sign. Hence the delayed replicas of $V_S(t)$ are added to and subtracted from the sum, resulting in oscillations.

So if the source and load reflection coefficients are of opposite sign, we will get ringing. On the other hand, if the source and load reflection coefficients are of the same sign, the load voltage will steadily build up to the steady-state value.

EXAMPLE

The question is: How do we remedy these signal integrity problems? The first simple scheme is *series matching*, where we place an additional resistor

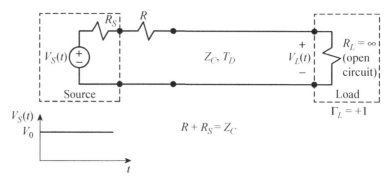

FIGURE 2.42. Series matching at the source end.

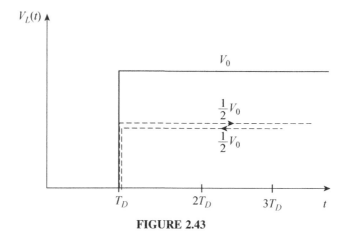

FIGURE 2.43

in series with the input to the line to bring the total source resistance to Z_C and therefore match the line at its source as illustrated in Fig. 2.42. Since the line is matched at the source, the voltage sent out initially is $\frac{1}{2}V_0$. So at the load we have the situation shown in Fig. 2.43. Since the source is matched at the source, there will be no reflections there, so that the load voltage will be brought up to the desired voltage of V_0.

The other matching scheme is parallel matching at the load as shown in Fig. 2.44. The voltage sent out initially is

$$V_{\text{init}} = \frac{Z_C}{R_S + Z_C} V_0$$

So the load voltage is as shown in Fig. 2.45.

There are two disadvantages to the parallel match scheme compared to the series match scheme. When the source and load are CMOS, their input impedances look capacitive and in the steady state appear as an open circuit and do not draw current.

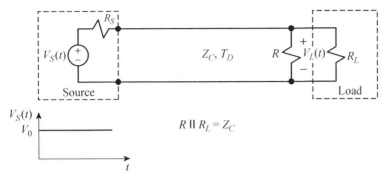

$$R \parallel R_L = Z_C$$

FIGURE 2.44. Parallel matching at the load end.

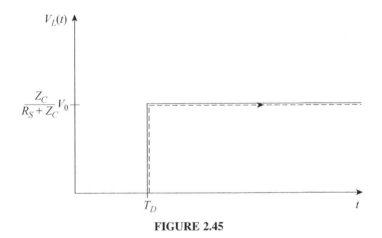

FIGURE 2.45

1. When the load is a CMOS gate, the R at the load draws current and dissipates unnecessary power.
2. The steady-state voltage at the load is

$$V_{L,\text{ss}} = \frac{R\|R_L = Z_C}{(R\|R_L = Z_C) + R_S} V_0$$

which is less than the desired V_0 that the series match achieves.

EXAMPLES OF SIGNAL INTEGRITY MATCHING

Regarding the problem of overshoot and undershoot, see Fig. 2.46.
    The PSPICE program is

```
EXAMPLE
VS 1 0 PWL(0 0 0.1N 5 5N 5 5.1N 0 10N 0)
RS 1 2 10
```

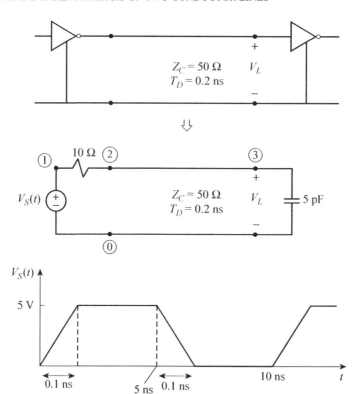

FIGURE 2.46. Example of overshoot and undershoot.

```
T 2 0 3 0 Z0=50 TD=0.2N
CL 3 0 5P
.TRAN 0.01N 10N 0 0.01N
.PROBE
.END
```

The load voltage is shown in Fig. 2.47.

EXAMPLE: SERIES MATCH AT THE SOURCE

Regarding the problem of series matching at the source, see Fig. 2.48.   The PSPICE program is

```
EXAMPLE
VS 1 0 PWL(0 0 0.1N 5 5N 5 5.1N 0 10N 0)
RS 1 2 10
```

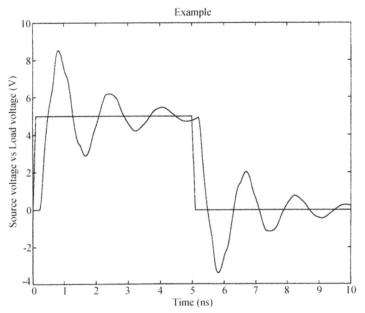

FIGURE 2.47. The load voltage.

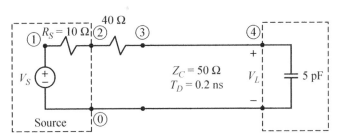

FIGURE 2.48. Series matching at the source.

```
R 2 3 40
T 3 0 4 0 Z0=50 TD=0.2N
CL 4 0 5P
.TRAN 0.01N 10N 0 0.01N
.PROBE
.END
```

The load voltage is shown in Fig. 2.49. Since the source is matched, the load capacitance "sees" a Thévenin impedance of Z_C, and therefore its voltage increases with a time constant of $Z_C C_L = 0.25$ ns.

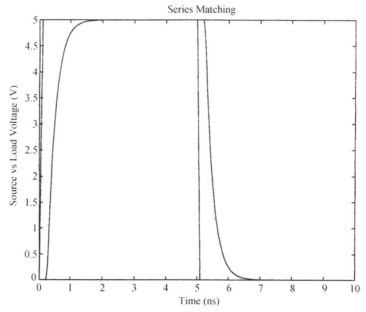

FIGURE 2.49. Load voltage due to series matching at the source.

EXAMPLE: PARALLEL MATCH AT THE LOAD

Regarding the problem of parallel matching at the load, see Fig. 2.50.   The PSPICE program is

```
EXAMPLE
VS 1 0 PWL(0 0 0.1N 5 5N 5 5.1N 0 10N 0)
RS 1 2 10
T 2 0 3 0 Z0=50 TD=0.2N
CL 3 0 5P
R 3 0 50
.TRAN 0.01N 10N 0 0.01N
.PROBE
.END
```

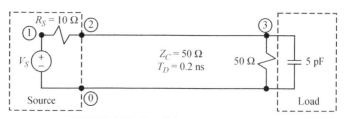

FIGURE 2.50. Parallel matching at the load.

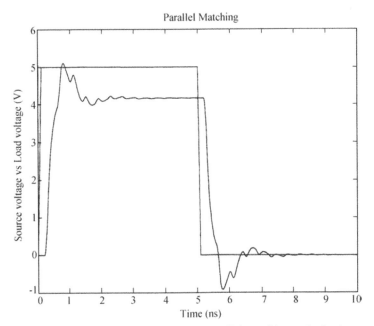

FIGURE 2.51. Load voltage due to parallel matching at the load.

The load voltage is shown in Fig. 2.51. What causes the "blip" in the load voltage? Isn't the line matched at the load? Note that the impedance of the capacitor is smaller than that of the resistor above the following frequency and hence the line is not matched at these frequency, components of $V_S(t)$ above:

$$f = \frac{1}{2\pi RC} = \frac{1}{2\pi \times 50 \times 5 \times 10^{-12}} = 637\,\text{MHz}$$

But since the source is not matched for these higher-frequency components of $V_S(t)$, these reflections at the load are rereflected at the source and keep bouncing back and forth along the line and cause the blip. The bandwidth of $V_S(t)$ is BW $= 1/\tau_r = 10\,\text{GHz}$, so there are a significant number of harmonics that are not matched by this scheme.

EXAMPLE: SERIES MATCHING AT THE SOURCE END AND PARALLEL MATCHING AT THE LOAD END

Since matching at the source and the load works well, one might wonder if matching at both ends might achieve both objectives. This is not the case, as this example shows. Since the source is matched, the voltage sent out initially

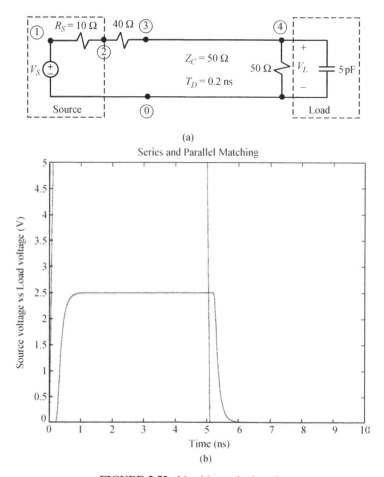

(a)

FIGURE 2.52. Matching at both ends.

is $V_0/2$, and since the load is not matched at certain high frequencies, there is no reflection to bring that up to the desired V_0. Since the source is matched (Fig. 2.52), these reflections at the load are absorbed at the source and cease being reflected at the source. The *steady-state* load voltage is $V_0/2 = 2.5$ V and logic errors will probably result!

2.9   BANDWIDTH AND SIGNAL INTEGRITY: WHEN DOES THE LINE NOT MATTER?

When do we need to match a transmission line to achieve signal integrity? Essentially, when the line is electrically short, we are reduced to lumped-circuit ideas, where we do not need to worry about distributed-parameter transmission-line effects. In other words, if the line is sufficiently short electrically, at the highest significant frequency of the source, we can consider it a lumped circuit. This requires that

$$\mathscr{L} \ll \frac{1}{10}\lambda = \frac{1}{10f_{max}}\frac{v}{} \tag{2.87}$$

where

$$BW = f_{max}$$
$$= \frac{1}{\tau_r} \tag{2.88}$$

Substituting gives the criterion for the line effects to be inconsequential as

$$\boxed{\tau_r \gg 10T_D} \tag{2.89}$$

Expressing the rise time in nanoseconds and the line length in inches and using an effective relative permittivity of $\varepsilon_r = (1 + 4.7)/2 = 2.85$ (for a microstripline) or $\varepsilon_r = 4.7$ (for a stripline) gives

$$\boxed{\begin{array}{lll} \tau_r(ns) > 1.43 & \mathscr{L}(in) & \text{microstrip} \\ \tau_r(ns) > 1.84 & \mathscr{L}(in) & \text{stripline} \end{array}} \tag{2.90}$$

To test this criterion we examine the following example.

EXAMPLE

For an example of how to avoid significant overshoot and undershoot, see Fig. 2.53. From the results in Fig. 2.53 we see that to avoid significant overshoot and undershoot, we require that (2.89) be satisfied.

2.10 EFFECT OF LINE DISCONTINUITIES

Line discontinuities occur in cables in the form of connectors and on PCBs in the form of disrupt changes in land widths as well as the presence of vias between layers in innerlayer PCBs. In this section we will see that line discontinuities also cause reflections and signal integrity problems in the same way that terminal discontiuities do.

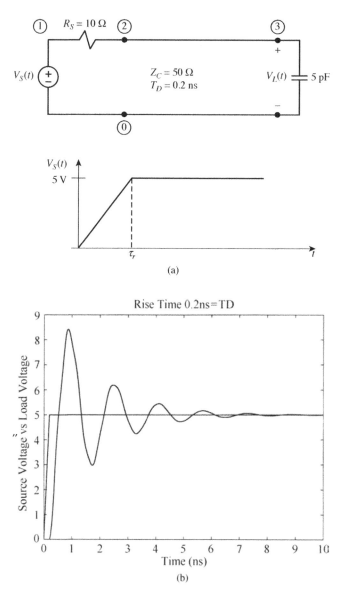

FIGURE 2.53

So far we have been considering *uniform lines* where the line cross section is uniform along its length; that is, its properties (radius of wires, separation of wires, land widths, and land separations) do not change along its length. Hence its per-unit-length parameters of inductance l and capacitance c do not depend on position z along its length. This greatly simplifies the solution of the transmission-line equations that contain them.

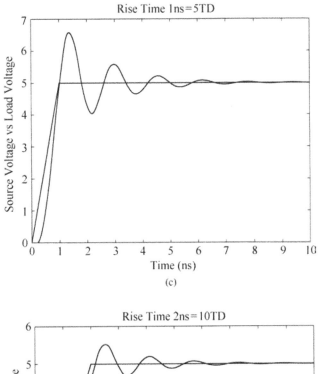

(c)

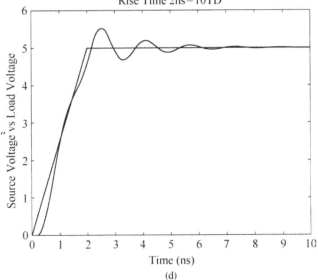

(d)

FIGURE 2.53. (*Continued*)

Suppose that the line changes its Z_C value at some point along its length, perhaps by widening out abruptly, as illustrated in Fig. 2.54. At the discontinuity, a wave coming from the left essentially sees the right half as a resistance Z_{C2} terminating the left half. So there is a voltage reflection

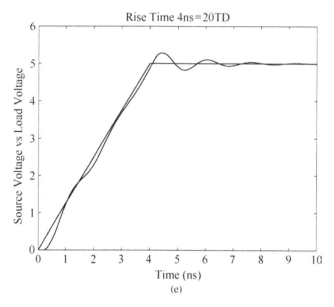

FIGURE 2.53. (*Continued*)

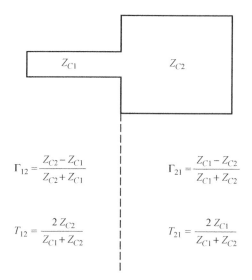

FIGURE 2.54. Changes in characteristic impedance by an abrupt change in land width on a PCB.

coefficient for voltage waves incident from the left side of

$$\boxed{\Gamma_{12} = \frac{Z_{C2} - Z_{C1}}{Z_{C2} + Z_{C1}}} \tag{2.91}$$

Similarly, for voltage waves traveling to the left in the second line, there is a voltage reflection coefficient on the right-hand side of the discontinuity of

$$\boxed{\Gamma_{21} = \frac{Z_{C1} - Z_{C2}}{Z_{C1} + Z_{C2}}} \tag{2.92}$$

In addition to the voltages reflected on each side of the discontinuity, there are also voltage waves *transmitted across the discontinuity*. This is shown in Fig. 2.55.

The *sum* of the incident and reflected voltages on one side of the discontinuity must equal the transmitted voltage on the other side of the discontinuity.

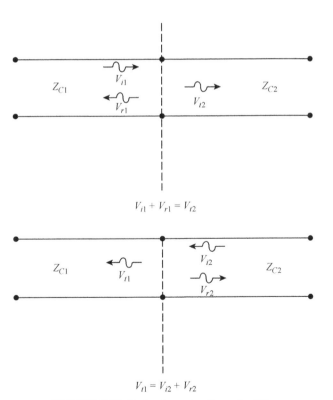

FIGURE 2.55. Reflections at a discontinuitity.

For voltage waves incident from the left side, we must have

$$v_{i1} + v_{r1} = v_{t2}$$
$$1 + \frac{v_{r1}}{v_{i1}} = \frac{v_{t2}}{v_{i1}} \tag{2.93}$$

or

$$\boxed{1 + \Gamma_{12} = T_{12}} \tag{2.94}$$

Hence we obtain the *voltage transmission coefficients* on either side of the boundary (exactly of the same form as the transmission coefficients for uniform plane waves incident normal to a boundary) as

$$\boxed{T_{12} = \frac{2Z_{C2}}{Z_{C2} + Z_{C1}}} \tag{2.95}$$

Similarly, for waves traveling to the left and incident on the discontinuity from the right side we must have

$$v_{i2} + v_{r2} = v_{t1} \tag{2.96}$$

So we obtain the voltage transmission coefficient

$$\boxed{T_{21} = \frac{2Z_{C1}}{Z_{C1} + Z_{C2}}} \tag{2.97}$$

and

$$\boxed{1 + \Gamma_{21} = T_{21}} \tag{2.98}$$

EXAMPLE

Consider the line discontinuity depicted in Fig. 2.56.

We simply sketch the waves on the line at points AA' and BB' and then add all the waves present on the graph (Fig. 2.57). The voltage sent out initially is

$$V_{\text{init}} = \frac{50}{50 + 50} V_S(t)$$
$$= 2.5 \text{ V}$$

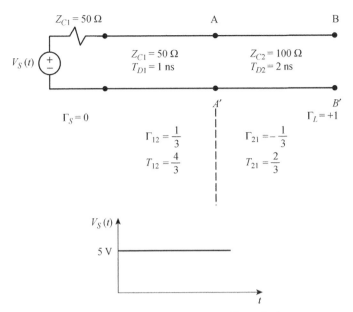

FIGURE 2.56. Example of a line discontinuity.

The PSPICE program is

```
EXAMPLE
VS 1 0 PWL(0 0 1P 5 100N 5)
RS 1 2 50
T 2 0 3 0 Z0=50 TD=1N
T 3 0 4 0 Z0=100 TD=2N
RL 4 0 1E8
.TRAN 0.1N 20N 0 0.1N
.PROBE
.END
```

The line voltages at the discontinuitities so obtained are sketched in Fig. 2.58.

2.11 DRIVING MULTIPLE LINES

It is very common to have one source, such as a digital clock driving numerous loads that are either in series or in parallel. Analysis of these by hand as we have done before is very involved and little design information can be obtained

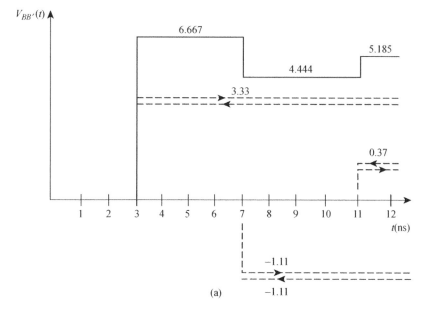

(a)

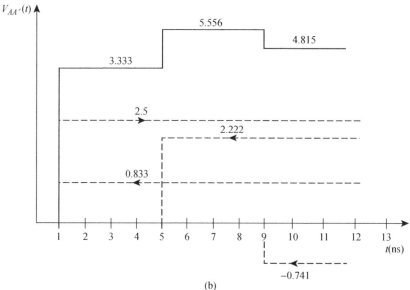

(b)

FIGURE 2.57. Line voltages for the discontuity in Fig. 2.56.

from this analysis. The simplest way to analyze these configurations and determine their performance is to model *each section* with PSPICE, interconnect these models in one PSPICE program, and then run the entire program to view the intermediate results to determine the resulting voltages and currents as we have done for the previous example in Fig. 2.56.

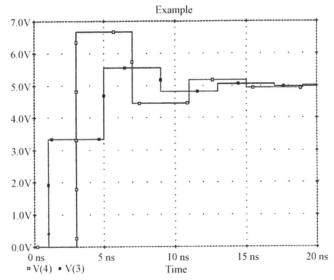

FIGURE 2.58. PSPICE solution for the line voltages of Fig. 2.56.

PROBLEMS

2.1 Show by direct substitution that the solution to the transmission-line equations

$$\frac{\partial V(z,t)}{\partial z} = -l\frac{\partial I(z,t)}{\partial t}$$

$$\frac{\partial I(z,t)}{\partial z} = -c\frac{\partial V(z,t)}{\partial t}$$

is

$$V(z,t) = \underbrace{V^+\left(t - \frac{z}{v}\right)}_{\substack{\text{forward } (+z) = \\ \text{traveling wave}}} + \underbrace{V^-\left(t + \frac{z}{v}\right)}_{\substack{\text{backward } (-z) = \\ \text{traveling wave}}}$$

and

$$I(z, t) = \underbrace{\frac{V^+\left(t - \dfrac{z}{v}\right)}{Z_C}}_{\substack{\text{forward } (+z) = \\ \text{traveling wave}}} - \underbrace{\frac{V^-\left(t + \dfrac{z}{v}\right)}{Z_C}}_{\substack{\text{backward } (-z) = \\ \text{traveling wave}}}$$

where the *characteristic impedance* is

$$Z_C = \sqrt{\frac{l}{c}} \quad \Omega$$

and the velocity of propagation is

$$v = \frac{1}{\sqrt{lc}} \quad \text{m/s}$$

2.2 Determine the exact and approximate values for the per-unit-length capacitance and inductance of a two-wire line (typical of ribbon cables used to interconnect electronic components) whose wires have radii of 7.5 mils (0.19 mm) and are separated by 50 mils (1.27 mm). [Exact: 14.8 pF/m and 0.75 µH/m; approximate: 14.6 pF/m and 0.759 µH/m]

2.3 Determine the exact and approximate values for the per-unit-length capacitance and inductance of one wire of radius 16 mils (0.406 mm) at a height of 100 mils (2.54 mm) above a ground plane. [Exact: 22.1 pF/m and 0.504 µH/m; approximate: 22.0 pF/m and 0.505 µH/m]

2.4 Determine the per-unit-length capacitance and inductance of a coaxial cable (RG-58U) where the inner wire has radius 16 mils (0.406 mm) and the shield has an inner radius of 58 mils (1.47 mm). The dielectric is polyethylene having a relative permittivity of 2.3. [99.2 pF/m and 0.258 µH/m]

2.5 Two bare No. 20 gauge (radius = 16 mils) wires are separated center to center by 50 mils. Determine the exact and approximate values of the per-unit-length capacitance and inductance. [Exact: 27.33 pF/m and

0.4065 µH/m; approximate: 24.38 pF/m and 0.4558 µH/m. Observe that the ratio of separation to wire radius is only 3.13, which is not sufficient for the approximate results to be valid.]

2.6 One bare No. 12 gauge (radius $= 40$ mils) is suspended at a height of 80 mils above its return path, which is a large ground plane. Determine the exact and approximate values of the per-unit-length capacitance and inductance. [Exact: 42.18 pF/m and 0.2634 µH/m; approximate: 40.07 pF/m and 0.2773 µH/m. The ratio of wire height above ground to wire radius is only 2.0, which is not sufficient for the approximate results to be valid, although the error is only about 5%.]

2.7 A typical coaxial cable is RG-6U, which has an interior No. 18 gauge (radius 20.15 mils) solid wire, an interior shield radius of 90 mils, and an inner insulation of foamed polyethylene having a relative permittivity of 1.45. Determine the per-unit-length capacitance, inductance, and the velocity of propagation relative to that of free space. [53.83 pF/m and 0.3 µH/m. The velocity of propagation relative to free space is $v/v_0 = 1/\sqrt{\varepsilon_r} = 0.83$.]

2.8 Determine the per-unit-length capacitance and inductance of a stripline with dimensions $s = 20$ mils (0.508 mm), $w = 5$ mils (0.127 mm), and $\varepsilon_r = 4.7$. [113.2 pF/m and 0.461 µH/m]

2.9 Determine the per-unit-length capacitance and inductance of a microstrip line with dimensions $h = 50$ mils (1.27 mm), $w = 5$ mils (0.127 mm), and $\varepsilon_r = 4.7$. [38.46 pF/m and 0.877 µH/m. The effective relative permittivity is $\varepsilon_r' = 3.034$.]

2.10 Determine the per-unit-length capacitance and inductance of a PCB with dimensions $s = 15$ mils (0.381 mm), $w = 15$ mils (0.381 mm), $h = 62$ mils (1.575 mils), and $\varepsilon_r = 4.7$. [38.53 pF/m and 0.804 µH/m. The effective relative permittivity is $\varepsilon_r' = 2.787$.]

2.11 Multilayer PCBs consist of layers of glass-epoxy material (FR-4) having a relative permittivity of 4.7 sandwiched between conducting planes. Conductors are buried midway between the conducting planes, which gives a structure resembling a stripline. Typical dimensions for multilayer PCBs are a plate separation of 10 mils and a conductor width of 5 mils. Determine the per-unit-length capacitance and inductance for this structure. [156.4 pF/m and 0.334 µH/m]

2.12 A microstrip line is constructed on a FR-4 board having a relative permittivity of 4.7. The board thickness is 64 mils and the land width is

10 mils. Determine the per-unit-length capacitance and inductance as well as the effective relative permittivity. [0.7873 μH/m, 43.46 pF/m and $\varepsilon_r' = 3.079$]

2.13 A PCB has land widths of 5 mils and an edge-to-edge separtaion of 5 mils. The board is glass-epoxy having a relative permittivity of 4.7 and a thickness of 47 mils. Determine the per-unit-length capacitance and inductance as well as the effective relative permittivity. [0.8038 μH/m, 39.06 pF/m, and $\varepsilon_r' = 2.825$]

2.14 Determine the characteristic impedance and velocity of propagation for the two-wire line in Problem 2.5. [122 Ω and 3×10^8 m/s]

2.15 Determine the characteristic impedance and velocity of propagation for the one-wire a ground plane line in Problem 2.6. [79 Ω and 3×10^8 m/s]

2.16 Determine the characteristic impedance and velocity of propagation for the coaxial line in Problem 2.7. [75 Ω and 2.5×10^8 m/s]

2.17 Determine the characteristic impedance and velocity of propagation for the coaxial line in Problem 2.11. [46 Ω and 1.38×10^8 m/s]

2.18 Determine the characteristic impedance and velocity of propagation for the microstrip line in Problem 2.12. [135 Ω and 1.71×10^8 m/s]

2.19 Determine the characteristic impedance and velocity of propagation for the PCB line in Problem 2.13. [143 Ω and 1.79×10^8 m/s]

2.20 Show how the per-unit-length inductance and capactiance can be found from the characteristic impedance and velocity of propagation. [$l = Z_C/v$ and $c = 1/vZ_C$]

2.21 Sketch the load voltage, $V(\mathcal{L}, t)$, and the input current to the line, $I(0, t)$, for a transmission line having a dc source $V_S = 10$ V, $R_S = 30\,\Omega$, $R_L = 150\,\Omega$, $Z_C = 50\,\Omega$, and $v = 3 \times 10^8$ m/s, with a total length of 30 cm for $0 < t < 10$ ns. The source is switched on the line at $t = 0$. What should these plots converge to in the steady state? [$V(\mathcal{L}, t)$, $0 < t < 1$ ns, 0 V, 1 ns $< t < 3$ ns, 9.375 V, 3 ns $< t < 5$ ns, 8.203 V, 5 ns $< t < 7$ ns, 8.35 V, 7 ns $< t < 9$ ns, 8.331 V, 9 ns $< t < 11$ ns, 8.334 V; steady state 8.333 V, and $I(0, t)$, $0 < t < 2$ ns, 0.125 A, 2 ns $< t < 4$ ns, 0.047 A, 4 ns $< t < 6$ ns, 0.057 A, 6 ns $< t < 8$ ns, 0.055 A, 8 ns $< t < 10$ ns, 0.056 A, steady state 0.056 A]

2.22 Sketch the load voltage, $V(\mathcal{L}, t)$, and the input voltage to the line, $V(0, t)$, for a transmission line a dc source $V_S = 5$ V, $R_S = 50\,\Omega$,

$R_L = \infty\,\Omega$, $Z_C = 100\,\Omega$, $v = 2.5 \times 10^8\,\text{m/s}$ and a total length of 50 cm for $0 < t < 20\,\text{ns}$. The source is switched on the line at $t = 0$. What should these plots converge to in the steady state? [$V(\mathcal{L}, t)$, $0 < t < 2\,\text{ns}$, $0\,\text{V}$, $2\,\text{ns} < t < 6\,\text{ns}$, $6.667\,\text{V}$, $6\,\text{ns} < t < 10\,\text{ns}$, $4.444\,\text{V}$ $10\,\text{ns} < t < 14\,\text{ns}$, $5.185\,\text{V}$, $14\,\text{ns} < t < 18\,\text{ns}$, $4.938\,\text{V}$, $18\,\text{ns} < t < 22\,\text{ns}$, $5.021\,\text{V}$, steady state $5\,\text{V}$; and $V(0, t)$, $0 < t < 4\,\text{ns}$, $3.33\,\text{V}$, $4\,\text{ns} < t < 8\,\text{ns}$, $5.556\,\text{V}$, $8\,\text{ns} < t < 12\,\text{ns}$, $4.815\,V$, $12\,\text{ns} < t < 16\,\text{ns}$, $5.062\,\text{V}$, $16\,\text{ns} < t < 20\,\text{ns}$, $4.979\,\text{V}$, steady state $5\,\text{V}$]

2.23 Sketch and the input voltage to the line, $V(0, t)$, and the load current, $I(\mathcal{L}, t)$, for for a transmission line having a dc source $V_S = 100\,\text{V}$, $R_S = 50\,\Omega$, $R_L = 0\,\Omega$, $Z_C = 100\,\Omega$, $v = 3 \times 10^8\,\text{m/s}$ having a total length of 300 m for $0 < t < 10\mu\,\text{s}$. What should these plots converge to in the steady state? [$V(0,t), 0 < t < 2\mu\text{s}, 66.67\,\text{V}$, $2\mu\text{s} < t < 4\mu\text{s}, 22.22\,\text{V}, 4\mu\text{s} < t < 6\mu\text{s}, 7.407\,V, 6\mu\text{s} < t < 8\mu\text{s}, 2.469$ $V, 8\mu\text{s} < t < 10\mu\text{s}, 0.823$, steady state $0\,\text{V}$; and $I(\mathcal{L},t), 0 < t < 1\mu\text{s}$, $0\,\text{V}, 1\mu\text{s} < t < 3\mu\text{s}, 1.33\,\text{A}, 3\mu\text{s} < t < 5\mu\text{s}, 1.778\,\text{A}, 5\mu\text{s} < t < 7\mu\text{s}$, $1.926\,\text{A}, 7\mu\text{s} < t < 9\mu\text{s}, 1.975\,\text{A},$ $9\mu\text{s} < t < 11\mu\text{s}, 1.992\,\text{A},$ steady state $2\,\text{A}$]

2.24 Sketch the input voltage to the line, $V(0, t)$, and the load voltage, $V(\mathcal{L}, t)$, for a transmission line having a pulse source of height $V_S = 30\,\text{V}$ and   duration   12 ns, $R_S = 200\,\Omega$, $R_L = 50\,\Omega$, $Z_C = 100\,\Omega$, $v = 2.5 \times 10^8\,\text{m/s}$, with a total length of 1 m for $0 < t < 32\,\text{ns}$. What should these plots converge to in the steady state? [$V(0,t), 0 < t < 8\,\text{ns}, 10$ $\text{V}, 8\,\text{ns} < t < 12\,\text{ns}, 5.556\,\text{V}, 12\,\text{ns} < t < 16\,\text{ns}$, $-4.444\,\text{V}$, $16\,\text{ns} < t < 20\,\text{ns}$, $-3.951\,\text{V}, 20\,\text{ns} < t < 24\,\text{ns}, 0.494\,\text{V}, 24$ $\text{ns} < t < 28\,\text{ns}, 0.439\,\text{V}$, $28\,\text{ns} < t$ $< 32\,\text{ns}$, -0.055 V. steady-state $0\,\text{V}$; and $V(\mathcal{L}, t), 0 < t < 4\,\text{ns}$, $0\,\text{V}, 4\,\text{ns} < t < 12\,\text{ns}, 6.667\,\text{V}, 12\,\text{ns}$ $< t < 16\,\text{ns}, 5.926\,\text{V}, 16\,\text{ns} < t < 20\,\text{ns}, -$ $0.741\,\text{V}, 20\,\text{ns} < t < 24\,\text{ns}$, $-0.658\,\text{V}, 24\,\text{ns} < t < 28\,\text{ns}, 0.082\,\text{V}, 28\,\text{ns} < t < 32\,\text{ns}, 0.073\,\text{V}$, steady-state $0\,\text{V}$]

2.25 A time-domain reflectometer (TDR) is an instrument used to determine properties of transmission lines. In particular, it can be used to detect the locations of imperfections such as breaks in a line. The instrument launches a pulse down the line and records the transit time for that pulse to be reflected at some discontinuity and to return to the line input. Suppose that a TDR having a source impedance of $50\,\Omega$ is attached to a $50\text{-}\Omega$ coaxial cable having some unknown length and load resistance. The dielectric of the cable is Teflon ($\varepsilon_r = 2.1$). The open-circuit voltage of the TDR is a pulse of duration $10\,\mu s$. If the recorded voltage at the input to the line is $V_S(t) = 100\,\text{V}$, $0 < t < 6\,\mu s$; $120\,\text{V}$, $6\,\mu s < t < 10\,\mu s$; $20\,\text{V}$, $10\,\mu s < t < 16\,\mu s$, $0\,\text{V}$, $16\,\mu s < t < 20\,\mu s$, determine (a) the length of the line and (b) the unknown load resistance. [$75\,\Omega$ and $621.059\,\text{m}$]

2.26 A 12-V battery $(R_S = 0)$ is attached to an unknown length of transmission lne that is terminated in a resistance. If the input current to the line for $6\,\mu s$ is $I(0, t) = 150\,\text{mA}$, $0 < t < 4.5\,\mu s$, $-10\,\text{mA}$, $4.5\,\text{mA} < t < 6\,\text{mA}$, determine (a) the line characteristic impedance, and (b) the unknown load resistance. [$Z_C = 80\,\Omega$, and $R_L = 262.9\,\Omega$]

2.27 Digital clock and data pulses should ideally consist of rectangular pulses. Actual clock and data pulses, however, resemble pulses having a trapezoidal shape with certain rise and fall times. Depending on the ratio of the rise and fall time to the one-way transit time of the transmission line, the voltage received may oscillate about the desired value, possibly causing a digital gate at that end to switch falsely to an undesired state and cause errors. Matching the line eliminates this problem because there are no reflections, but matching cannot always be accomplished. To investigate this problem, consider a line connecting two CMOS gates. The driver gate is assumed to have zero source resistance $(R_S = 0)$, and the open-circuit voltage is a ramp waveform (simulating the leading edge of the clock/data pulse) given by $V_S(t) = 0$ for $t < 0$, $V_S(t) = 5(t/\tau_r)\,\text{V}$ for $0 \le t \le \tau_r$, and $V_S(t) = 5\,\text{V}$ for $t \ge \tau_r$, where τ_r is the pulse rise time. The input to a CMOS gate (the load on the line here) can be modeled as a capacitance of some 5 to 15pF. However, to simplify the problem we will assume that the input to the load CMOS gate is an open circuit $R_L = \infty$. Sketch the load voltage of the line (the input voltage to the load CMOS gate) for line lengths having one-way transit times T_D such that (a) $\tau_r = T_D/10$, (b) $\tau_r = 2T_D$, (c) $\tau_r = 3T_D$, (d) $\tau_r = 4T_D$. This example shows that to avoid problems resulting from mismatch, one should choose line lengths short enough such that $T_D \ll \tau_r$, that is, the line one-way delay is much less than the rise time of the clock/data pulses being carried by the line.

2.28 Highly mismatched lines in digital products can cause what appear to be ringing on the signal output from the line. This is often referred to as overshoot or undershoot and can cause digital logic errors. To simulate this we investigate a problem in which two CMOS gates are connected by a transmission line where the driver gate has $R_S = Z_C/5$ and the load gate has $R_L = 5Z_C$. A 5-V step function voltage of the first gate is applied. Sketch the output voltage of the line (the input voltage to the load CMOS gate) for $0 < t < 9T_D$. [$0 < t < T_D, 0\,\mathrm{V}, T_D < t < 3T_D, 6.944\,\mathrm{V}, 3T_D < t < 5T_D, 3.858\,\mathrm{V}, \; 5T_D < t < 7T_D, 5.23\,\mathrm{V}, 7T_D < t < 9T_D, 4.62\,\mathrm{V}$, steady state is 5 V]

2.29 A transmission line of total length 200 m and velocity of propagation of $v = 2 \times 10^8\,\mathrm{m/s}$ has $Z_C = 50\,\Omega$ and $R_L = 20\,\Omega$. It is driven by a source having $R_S = 100\,\Omega$ and an open-circuit voltage that is a rectangular pulse of 6 V magnitude and 3 μs duration. Sketch the input current to the line for a total time of 5 μs.

2.30 Confirm the results of Problem 2.21 using SPICE (PSPICE).

2.31 Confirm the results of Problem 2.22 using SPICE (PSPICE).

2.32 Confirm the results of Problem 2.23 using SPICE (PSPICE).

2.33 Confirm the results of Problem 2.24 using SPICE (PSPICE).

2.34 Confirm the results of Problem 2.25 using SPICE (PSPICE).

2.35 Confirm the results of Problem 2.26 using SPICE (PSPICE).

2.36 Confirm the results of Problem 2.27 using SPICE (PSPICE).

2.37 Confirm the results of Problem 2.28 using SPICE (PSPICE).

2.38 Confirm the results of Problem 2.29 using SPICE (PSPICE).

2.39 One of the important advantages in using SPICE to solve transmission-line problems is that it will readily give the solution for problems that would be difficult to solve by hand. For example, consider the case of two CMOS inverter gates connected by a 5-cm length of 100-Ω transmission line. The output of the driver gate is represented by a ramp waveform voltage rising from 0 to 5 V in 1ns and a 30-Ω internal source resistance. The receiving gate is represented at its input by 10 pF. Because of the capacitive load, this would be a difficult problem to solve by hand. Use SPICE (PSPICE) to plot the output voltage of the line, $V_L(t)$, for $0 < t < 10$ ns. Observe in the solution that this output voltage varies rather drastically about the desired 5-V level going from 4.2 to7 V before it stabilizes to 5 V well after 10 ns. Hence there is the distinct possibility of logic errors and something must be done.

2.40   Determine and sketch the voltages at the input and output of the line shown in Fig. P.2.40. Check your results using PSPICE.

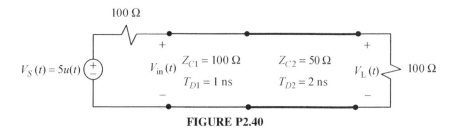

FIGURE P2.40

2.41   Determine and sketch the load voltage for the line shown in Fig. P.2.41.

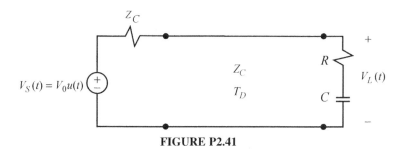

FIGURE P2.41

3

FREQUENCY-DOMAIN ANALYSIS OF TWO-CONDUCTOR LINES

In this chapter we examine the response of a two-conductor transmission line to a single-frequency sinusoidal source $V_S(t) = V_S \cos(\omega t + \phi)$ or $V_S(t) = V_S \sin(\omega t + \phi)$. We assume that the source has been applied for a sufficient length of time so that the line response voltages and currents have reached *steady state*. In other words, we examine the *phasor* response of the line, as illustrated in Fig. 3.1. So we replace the time-domain source with the phasor source:

$$\left.\begin{array}{c} V_S(t) = V_S \sin(\omega t + \phi) \\ V_S(t) = V_S \cos(\omega t + \phi) \end{array}\right\} \Rightarrow \underbrace{V_S \angle \phi}_{\hat{V}_S}\, e^{j\omega t} \tag{3.1}$$

The line voltages and currents are (complex-valued) phasors with a magnitude and an angle that we denote with carets:

$$V(z,t) \Rightarrow \underbrace{V \angle \theta_V}_{\hat{V}(z)} e^{j\omega t} \tag{3.2a}$$

$$I(z,t) \Rightarrow \underbrace{I \angle \theta_I}_{\hat{I}(z)} e^{j\omega t} \tag{3.2b}$$

Transmission Lines in Digital and Analog Electronic Systems: Signal Integrity and Crosstalk, By Clayton R. Paul
Copyright © 2010 John Wiley & Sons, Inc.

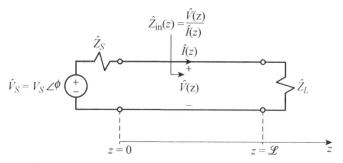

FIGURE 3.1. Phasor response of the two-conductor line.

3.1   THE TRANSMISSION-LINE EQUATIONS FOR SINUSOIDAL STEADY-STATE EXCITATION OF THE LINE

The time-domain derivatives are replaced with $j\omega$:

$$\frac{\partial}{\partial t} \Rightarrow j\omega \tag{3.3}$$

where the radian frequency of the source is $\omega = 2\pi f\,(\text{rad/s})$ and f is its cyclic frequency in hertz. The (phasor) transmission-line equations for a *lossless* line become

$$\boxed{\frac{d\hat{V}(z)}{dz} = -j\omega l\,\hat{I}(z)} \tag{3.4a}$$

$$\boxed{\frac{d\hat{I}(z)}{dz} = -j\omega c\,\hat{V}(z)} \tag{3.4b}$$

The uncoupled second-order transmission-line equations become

$$\boxed{\frac{d^2\hat{V}(z)}{dz^2} + \underbrace{\omega^2 lc}_{\beta^2}\,\hat{V}(z) = 0} \tag{3.5a}$$

$$\boxed{\frac{d^2\hat{I}(z)}{dz^2} + \underbrace{\omega^2 lc}_{\beta^2}\,\hat{I}(z) = 0} \tag{3.5b}$$

3.2   THE GENERAL SOLUTION FOR THE TERMINAL VOLTAGES AND CURRENTS

The general phasor solutions to these phasor equations are easily obtained as

$$\hat{V}(z) = \hat{V}^{+} e^{-j\beta z} + \hat{V}^{-} e^{j\beta z} \tag{3.6a}$$

$$\hat{I}(z) = \frac{\hat{V}^{+}}{Z_C} e^{-j\beta z} - \frac{\hat{V}^{-}}{Z_C} e^{j\beta z} \tag{3.6b}$$

where the *phase constant* is

$$\begin{aligned} \beta &= \omega\sqrt{lc} \\ &= \frac{\omega}{v} \qquad \text{rad/m} \end{aligned} \tag{3.7}$$

The time-domain solutions are

$$V(z,t) = V^{+} \cos(\omega t - \beta z + \theta^{+}) + V^{-} \cos(\omega t + \beta z + \theta^{-}) \tag{3.8a}$$

$$I(z,t) = \frac{V^{+}}{Z_C} \cos(\omega t - \beta z + \theta^{+}) - \frac{V^{-}}{Z_C} \cos(\omega t + \beta z + \theta^{-}) \tag{3.8b}$$

and the undetermined constants in the general solution, $\hat{V}^{+} = V^{+} \angle\theta^{+}$ and $\hat{V}^{-} = V^{-} \angle\theta^{-}$, have a magnitude and a phase.

3.3   THE VOLTAGE REFLECTION COEFFICIENT AND INPUT IMPEDANCE TO THE LINE

The voltage reflection coefficient is defined as

$$\begin{aligned} \hat{\Gamma}(z) &= \frac{\hat{V}^{-} e^{j\beta z}}{\hat{V}^{+} e^{-j\beta z}} \\ &= \frac{\hat{V}^{-}}{\hat{V}^{+}} e^{j2\beta z} \end{aligned} \tag{3.9}$$

This is a general voltage reflection coefficient at any z along the line. Evaluating this at the load, $z = \mathcal{L}$, gives

$$\hat{\Gamma}_L = \frac{\hat{V}^-}{\hat{V}^+} e^{j2\beta \mathcal{L}} \tag{3.10}$$

where the voltage reflection coefficient at the load is, as before,

$$\boxed{\hat{\Gamma}_L = \frac{\hat{Z}_L - Z_C}{\hat{Z}_L + Z_C}} \tag{3.11}$$

The voltage reflection coefficient at any point along the line can then be written in terms of the load reflection coeffcient as

$$\boxed{\hat{\Gamma}(z) = \hat{\Gamma}_L e^{j2\beta (z - \mathcal{L})}} \tag{3.12}$$

The general solutions at any point on the line can be written in terms of the voltage reflection coefficient there as

$$\boxed{\hat{V}(z) = \hat{V}^+ e^{-j\beta z} \left[1 + \hat{\Gamma}(z)\right]} \tag{3.13a}$$

$$\boxed{\hat{I}(z) = \frac{\hat{V}^+}{Z_C} e^{-j\beta z} \left[1 - \hat{\Gamma}(z)\right]} \tag{3.13a}$$

Substituting the explicit relation for the reflection coefficient in terms of the load reflection given in (3.12) gives

$$\boxed{\hat{V}(z) = \hat{V}^+ e^{-j\beta z} \left[1 + \hat{\Gamma}_L e^{j2\beta (z - \mathcal{L})}\right]} \tag{3.14a}$$

$$\boxed{\hat{I}(z) = \frac{\hat{V}^+}{Z_C} e^{-j\beta z} \left[1 - \hat{\Gamma}_L e^{j2\beta (z - \mathcal{L})}\right]} \tag{3.14b}$$

The input impedance at any point on the line is defined as

$$
\begin{aligned}
\hat{Z}_{\text{in}}(z) &= \frac{\hat{V}(z)}{\hat{I}(z)} \\
&= Z_C \left[\frac{1 + \hat{\Gamma}(z)}{1 - \hat{\Gamma}(z)} \right]
\end{aligned}
$$

(3.15)

The reflection coefficients at the input and at the load are related by

$$
\hat{\Gamma}(0) = \hat{\Gamma}_L e^{-j2\beta\mathscr{L}}
$$

(3.16)

Hence the input impedance to a line of length $\mathscr{L}$ is

$$
\begin{aligned}
\hat{Z}_{\text{in}} &= Z_C \left[\frac{1 + \hat{\Gamma}(0)}{1 - \hat{\Gamma}(0)} \right] \\
&= Z_C \left[\frac{1 + \hat{\Gamma}_L e^{-j2\beta\,\mathscr{L}}}{1 - \hat{\Gamma}_L e^{-j2\beta\,\mathscr{L}}} \right]
\end{aligned}
$$

(3.17)

where $\hat{Z}_{\text{in}} \equiv \hat{Z}_{\text{in}}(0)$.

3.4 THE SOLUTION FOR THE TERMINAL VOLTAGES AND CURRENTS

The phasor solutions for the terminal voltages and currents are

$$
\hat{V}(0) = \hat{V}^+ \left[1 + \hat{\Gamma}(0) \right]
$$

(3.18a)

$$
\hat{I}(0) = \frac{\hat{V}^+}{Z_C} \left[1 - \hat{\Gamma}(0) \right]
$$

(3.18b)

and

$$
\hat{V}(\mathscr{L}) = \hat{V}^+ e^{-j\beta\,\mathscr{L}} \left[1 + \hat{\Gamma}_L \right]
$$

(3.19a)

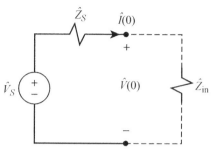

FIGURE 3.2. Equivalent input circuit to the phasor line.

$$\hat{I}(\mathscr{L}) = \frac{\hat{V}^+}{Z_C} e^{-j\beta\mathscr{L}} \left[1 - \hat{\Gamma}_L\right] \qquad (3.19b)$$

We only need to determine $\hat{V}^+$. The input to the phasor line appears as shown in Fig. 3.2. Hence, by voltage division we can determine the phasor input voltage to the line as

$$\hat{V}(0) = \frac{\hat{Z}_{in}}{\hat{Z}_S + \hat{Z}_{in}} \hat{V}_S \qquad (3.20)$$

and the undetermined constant is determined as

$$\hat{V}^+ = \frac{\hat{V}(0)}{1 + \hat{\Gamma}(0)} \qquad (3.21)$$

EXAMPLE

Consider the circuit shown in Fig. 3.3. How long is the line in wavelengths?

$$\mathscr{L} = k\frac{v}{f}$$
$$= 1.35\lambda$$

The load voltage reflection coefficient is

$$\hat{\Gamma}_L = \frac{\hat{Z}_L - Z_C}{\hat{Z}_L + Z_C}$$
$$= \frac{50 + j200}{150 + j200}$$
$$= 0.82\angle 22.83°$$

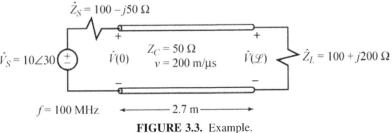

FIGURE 3.3. Example.

and the input voltage reflection coefficient is

$$\hat{\Gamma}(0) = \hat{\Gamma}_L e^{-j2\beta \mathscr{L}}$$

But

$$-2\beta \mathscr{L} = -4\pi f \frac{\mathscr{L}}{v}$$

$$= -4\pi \frac{\mathscr{L}}{\lambda}$$

Therefore,

$$\hat{\Gamma}(0) = \hat{\Gamma}_L e^{-j2\beta \mathscr{L}}$$

$$= 0.82\angle -949.17°$$

and the input impedance to the line is

$$\hat{Z}_{in} = Z_C \left[\frac{1 + \hat{\Gamma}(0)}{1 - \hat{\Gamma}(0)} \right]$$

$$= 23.35\angle 75.62°$$

Therefore, the input voltage to the line is

$$\hat{V}(0) = \frac{\hat{Z}_{in}}{\hat{Z}_S + \hat{Z}_{in}} \hat{V}_S$$

$$= 2.14\angle 120.13°$$

The undetermined constant is

$$\hat{V}^+ = \frac{\hat{V}(0)}{1 + \hat{\Gamma}(0)}$$

$$= 2.75\angle 66.58°$$

The input and load phasor voltages are

$$\hat{V}(0) = \hat{V}^+ \left[1 + \hat{\Gamma}(0)\right]$$

$$= 2.14\angle 120.13°$$

$$\hat{V}(\mathcal{L}) = \hat{V}^+ e^{-j\beta \mathcal{L}} \left[1 + \hat{\Gamma}_L\right]$$

$$= 4.93\angle - 409.12°$$

Hence the time-domain terminal voltages are

$$V(0, t) = 2.14\cos(6.28 \times 10^8 t + 120.13°)$$

$$V(\mathcal{L}, t) = 4.93\cos(6.28 \times 10^8 t - 49.12°)$$

3.5   THE SPICE SOLUTION

There are three changes from the time-domain use of the PSPICE for the phasor use of PSPICE.

1. The specification of the voltage source is changed to

   VS N1 N2 AC mag phase

2. The .TRAN line needs to be changed to

   .AC DEC 1 f f

   where f is the frequency of the source, and
3. The output is obtained as

   .PRINT AC VM(NX) VP(NX)

   where VM is the magnitude of the phasor voltage at node NX and VP is the phase.

EXAMPLE

To solve the previous phasor example using PSPICE, we must generate equivalent circuits to represent the complex impedances $\hat{Z}_S$ and $\hat{Z}_L$ using

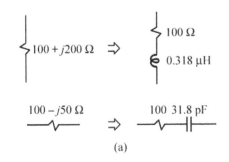

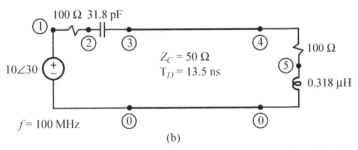

FIGURE 3.4. Lumped-equivalent circuit for the phasor circuit of Fig. 3.3.

lumped-circuit elements such as R's, L's, and C's, as illustrated in Fig. 3.4, that give the same complex impedances at the source frequency.

The PSPICE program is

```
EXAMPLE
VS 1 0 AC 10 30
RS 1 2 100
CS 2 3 31.8P
T 3 0 4 0 Z0=50 TD=13.5N
RL 4 5 100
LL 5 0 0.318U
.AC DEC 1 1E8 1E8
.PRINT AC VM(3) VP(3) VM(4) VP(4)
+ IM(CS) IP(CS)
.END
```

giving

$$V(3) = 2.136\angle 120.1° \quad V(4) = 4.926\angle -49.1°$$

and the input impedance to the line is computed as

$$\hat{Z}_{\text{in}} = \frac{\text{VM}(3)}{\text{IM}(\text{CS})} \angle[\text{VP}(3) - \text{IP}(\text{CS})] = 23.349\angle75.56°$$

as computed by hand.

3.6 VOLTAGE AND CURRENT AS A FUNCTION OF POSITION ON THE LINE

Previously we determined the voltage and current for sinusoidal excitation only at the endpoints of the line. *How do they vary along the line?* The voltage and current along the line are

$$\hat{V}(z) = \hat{V}^+ e^{-j\beta z}\left[1 + \hat{\Gamma}_L e^{j2\beta (z - \mathcal{L})}\right] \tag{3.22a}$$

$$\hat{I}(z) = \frac{\hat{V}^+}{Z_C} e^{-j\beta z}\left[1 - \hat{\Gamma}_L e^{j2\beta (z - \mathcal{L})}\right] \tag{3.22b}$$

We will plot the magnitude of these for distances $d = \mathcal{L} - z$ away from the load. Taking the magnitudes of the phasor voltage and current gives

$$\left|\hat{V}(d = \mathcal{L} - z)\right| = \left|\hat{V}^+\right|\left|1 + \hat{\Gamma}_L e^{-j2\beta d}\right| \tag{3.23a}$$

and

$$\left|\hat{I}(d = \mathcal{L} - z)\right| = \frac{\left|\hat{V}^+\right|}{Z_C}\left|1 - \hat{\Gamma}_L e^{-j2\beta d}\right| \tag{3.23b}$$

Three important cases of special interest that we will investigate are shown in Fig. 3.5: (1) the load is a short circuit, $\hat{Z}_L = 0$; (2) the load is an open circuit, $\hat{Z}_L = \infty$; and (3) the load is matched, $\hat{Z}_L = Z_C$.

For the case where the load is a short circuit, $\hat{Z}_L = 0$, the load reflection coefficient is $\hat{\Gamma}_L = -1$ and the equations reduce to

$$\left|\hat{V}(d)\right| = \left|\hat{V}^+\right|\left|1 - e^{-j2\beta d}\right|$$

$$= \left|\hat{V}^+\right|\underbrace{\left|e^{-j\beta d}\right|}_{1}\underbrace{\left|e^{j\beta d} - e^{-j\beta d}\right|}_{|2j\sin\beta d|}$$

$$\propto \left|\sin\beta d\right|$$

$$= \left|\sin\left(2\pi\frac{d}{\lambda}\right)\right|$$

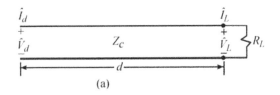

(a)

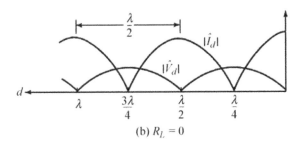

(b) $R_L = 0$

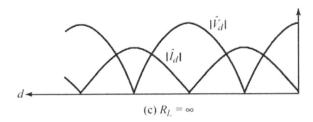

(c) $R_L = \infty$

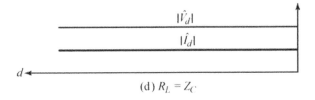

(d) $R_L = Z_C$

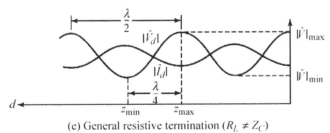

(e) General resistive termination $(R_L \neq Z_C)$

FIGURE 3.5

and

$$
\begin{aligned}
\left|\hat{I}(d)\right| &= \frac{\left|\hat{V}^{+}\right|}{Z_C}\left|1+e^{-j2\beta\,d}\right| \\
&= \frac{\left|\hat{V}^{+}\right|}{Z_C}\underbrace{\left|e^{-j\beta\,d}\right|}_{1}\underbrace{\left|e^{j\beta\,d}+e^{-j\beta\,d}\right|}_{|2\cos\beta d|} \\
&\propto \left|\cos\beta d\right| \\
&= \left|\cos\left(2\pi\frac{d}{\lambda}\right)\right|
\end{aligned}
$$

The voltage is zero at points away from the short-circuit load where d is a multiple of a half-wavelength, and is a maximum at points where d is an odd multiple of a quarter-wavelength.

For the case where the load is an open circuit, $\hat{Z}_L = \infty$, and the load reflection coefficient is $\hat{\Gamma}_L = 1$, the equations reduce to

$$
\begin{aligned}
\left|\hat{V}(d)\right| &= \left|\hat{V}^{+}\right|\left|1+e^{-j2\beta\,d}\right| \\
&\propto \left|\cos\beta d\right| \\
&= \left|\cos\left(2\pi\frac{d}{\lambda}\right)\right|
\end{aligned}
$$

and

$$
\begin{aligned}
\left|\hat{I}(d)\right| &= \frac{\left|\hat{V}^{+}\right|}{Z_C}\left|1-e^{-j2\beta\,d}\right| \\
&\propto \left|\sin\beta d\right| \\
&= \left|\sin\left(2\pi\frac{d}{\lambda}\right)\right|
\end{aligned}
$$

The current is zero at points away from the short-circuit load where d is a multiple of a half-wavelength, and is a maximum at points where d is an odd multiple of a quarter-wavelength.

For a matched line there is no variation in the voltage and current along the line. That's the advantage of being *matched*, $R_L = Z_C$.

Notice that whatever the value of the load impedance, the voltage (current) maximum and adjacent voltage (current) minimum are separated by $\lambda/4$; the corresponding voltage or current points are separated by $\lambda/2$.

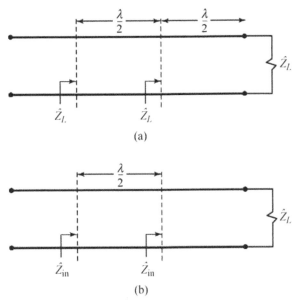

FIGURE 3.6. Replication of the input impedance for line lengths that are a multiple of a half-wavelength.

From (3.17), the input impedance to the line replicates for line lengths that are a multiple of $\lambda/2$ since $2\beta\mathcal{L} = 4\pi(\mathcal{L}/\lambda)$ and $e^{\pm j2\pi} = 1$ (see Fig. 3.6).

3.7   MATCHING AND VSWR

If the line is not matched, how do we *quantitatively* judge the *degree of the mismatch*? The answer is the *voltage standing-wave ratio* (VSWR), defined as

$$\text{VSWR} = \frac{\left|\hat{V}\right|_{\text{max}}}{\left|\hat{V}\right|_{\text{min}}} \tag{3.24}$$

A maximum and the adjacent minimum are separated by exactly $\lambda/4$.

For a short-circuit load, $\hat{Z}_L = 0$, the voltage minimum is zero, so the VSWR is infinite. Also, for an open-circuit load, $\hat{Z}_L = \infty$, the voltage minimum is zero, so the VSWR is infinite. Thus

$$VSWR = \begin{cases} \infty & \hat{Z}_L = 0 \\ \infty & \hat{Z}_L = \infty \end{cases}$$

but the VSWR is unity for a matched load:

$$VSWR = 1 \qquad \hat{Z}_L = Z_C$$

The VSWR must therefore lie between these two bounds:

$$\boxed{1 \le VSWR < \infty} \tag{3.25}$$

Industry considers a line to be *matched* if VSWR < 1.2. The VSWR can be calculated directly by taking the ratios of the *magnitudes* of the maximum and minimum of (3.14a):

$$\boxed{VSWR = \frac{1 + |\hat{\Gamma}_L|}{1 - |\hat{\Gamma}_L|}} \tag{3.26}$$

3.8   POWER FLOW ON THE LINE

At any point on the line, the power flow to the right, as illustrated in Fig. 3.7, is

$$P_{AV}(z) = \frac{1}{2} \mathrm{Re}\left\{ \hat{V}(z)\hat{I}^*(z) \right\} \tag{3.27}$$

where * denotes the complex conjugate. Hence

$$P_{AV}(z) = \frac{1}{2}\mathrm{Re}\left\{ \hat{V}(z)\hat{I}^*(z) \right\}$$

$$= \frac{1}{2}\mathrm{Re}\left\{ \hat{V}^+ e^{-j\beta z}\left[1 + \hat{\Gamma}_L e^{j2\beta(z-\mathscr{L})}\right]\frac{\hat{V}^{+*}}{Z_C}e^{j\beta z}\left[1 - \hat{\Gamma}_L^* e^{-j2\beta(z-\mathscr{L})}\right] \right\}$$

$$= \frac{1}{2}\mathrm{Re}\left\{ \frac{\hat{V}^+ \hat{V}^{+*}}{Z_C}\left[1 + \underbrace{\left(\hat{\Gamma}_L e^{j2\beta(z-\mathscr{L})} - \hat{\Gamma}_L^* e^{-j2\beta(z-\mathscr{L})}\right)}_{\text{imaginary}} - \hat{\Gamma}_L\hat{\Gamma}_L^*\right] \right\}$$

$$= \frac{\left|\hat{V}^+\right|^2}{2Z_C}\left[1 - |\hat{\Gamma}_L|^2\right]$$

$$\tag{3.28}$$

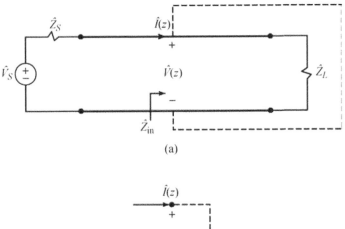

(a)

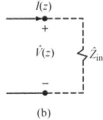

(b)

FIGURE 3.7. Power flow to the right on the line.

But the power in the forward-traveling wave is

$$P_{AV}^{+}(z) = \frac{1}{2}\text{Re}\left\{ \left(\hat{V}^{+}e^{-j\beta z}\right)\left(\frac{\hat{V}^{+*}}{Z_C}e^{j\beta z}\right)\right\}$$
$$= \frac{\left|\hat{V}^{+}\right|^{2}}{2Z_C} \tag{3.29}$$

and the power in the backward-traveling wave is

$$P_{AV}^{-}(z) = \frac{1}{2}\text{Re}\left\{ \left(\hat{V}^{-}e^{j\beta z}\right)\left(-\frac{\hat{V}^{-*}}{Z_C}e^{-j\beta z}\right)\right\}$$
$$= -\frac{\left|\hat{V}^{+}\right|^{2}}{2Z_C}\left|\hat{\Gamma}_{L}\right|^{2} \tag{3.30}$$

So the portion of the incident power that is reflected at the load is $\left|\hat{\Gamma}_{L}\right|^{2} \times 100\%$.

If the line is *lossless*, the total power delivered to the load is equal to the total power delivered to the input to the line! (see Fig. 3.8.)

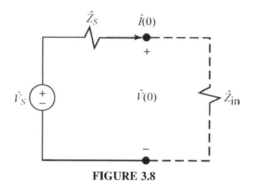

FIGURE 3.8

So to determine the power delivered to the load for a *lossless* line, you simply need to determine the power delivered to the input to the line.

$$P_{\text{AV, to load}} = P_{\text{AV, to line input}} = \frac{1}{2} \frac{|\hat{V}(0)|^2}{|\hat{Z}_{\text{in}}|} \cos(\theta_{Z_{\text{in}}}) \qquad (3.31)$$

EXAMPLE

In the preceding example we obtained

$$\hat{V}(0) = \hat{V}^+ \left[1 + \hat{\Gamma}(0)\right]$$
$$= 2.14\angle 120.13°$$
$$\hat{V}(\mathcal{L}) = \hat{V}^+ e^{-j\beta \mathcal{L}}(1 + \hat{\Gamma}_L)$$
$$= 4.93\angle -409.12°$$
$$\hat{Z}_{\text{in}} = Z_C \left[\frac{1 + \hat{\Gamma}(0)}{1 - \hat{\Gamma}(0)}\right]$$
$$= 23.35\angle 75.62°$$

Hence the power delivered to the line input and the load (since the line is lossless) is

$$P_{\text{AV, to input}} = \frac{1}{2} \frac{|\hat{V}(0)|^2}{|\hat{Z}_{\text{in}}|} \cos(\theta_{Z_{\text{in}}})$$
$$= \frac{1}{2} \frac{(2.14)^2}{23.35} \cos(75.62°)$$
$$= 24.3 \text{ mW}$$

and (directly)

$$P_{AV, \text{ to load}} = \frac{1}{2} \frac{|\hat{V}(\mathcal{L})|^2}{|\hat{Z}_L|} \cos(\theta_{Z_L})$$

$$= \frac{1}{2} \frac{(4.93)^2}{223.6} \cos(63.43°)$$

$$= 24.3 \text{ mW}$$

where $\hat{Z}_L = 100 + j200 = 223.6\angle 63.43°$. The VSWR is $(\hat{\Gamma}_L = 0.82\angle 22.83°)$

$$\text{VSWR} = \frac{1 + |\hat{\Gamma}_L|}{1 - |\hat{\Gamma}_L|}$$

$$= \frac{1 + 0.82}{1 - 0.82}$$

$$= 10.1$$

3.9   ALTERNATIVE FORMS OF THE RESULTS

The previous results for the line voltages and currents and the input impedance are the simplest to use. However, alternative formulas can be obtained (you should derive these):

$$\hat{V}(z) = \frac{1 + \hat{\Gamma}_L e^{-j2\beta\mathcal{L}} e^{j2\beta z}}{1 - \hat{\Gamma}_S \hat{\Gamma}_L e^{-j2\beta\mathcal{L}}} \frac{Z_C}{\hat{Z}_S + Z_C} \hat{V}_S e^{-j\beta z} \qquad (3.32)$$

$$\hat{I}(z) = \frac{1 - \hat{\Gamma}_L e^{-j2\beta\mathcal{L}} e^{j2\beta z}}{1 - \hat{\Gamma}_S \hat{\Gamma}_L e^{-j2\beta\mathcal{L}}} \frac{1}{\hat{Z}_S + Z_C} \hat{V}_S e^{-j\beta z} \qquad (3.33)$$

$$\hat{Z}_{\text{in}} = Z_C \frac{\hat{Z}_L + jZ_C \tan\beta\mathcal{L}}{Z_C + j\hat{Z}_L \tan\beta\mathcal{L}} \qquad (3.34)$$

These give considerable insight into the line behavior. You should check these for matched lines, $\hat{Z}_L = Z_C$.

3.10   THE SMITH CHART

The Smith chart is a *nomograph* for plotting and determining the input impedance to a transmission line. It can also be used to determine many other properties of the line *without solving any equations.* You see the use of the Smith chart throughout the trade journals. Microwave engineers "think in terms of the Smith chart"!

The input impedance to a transmission line of length $\mathscr{L}$ is

$$\hat{Z}_{in} = Z_C \left[\frac{1 + \hat{\Gamma}_{in}}{1 - \hat{\Gamma}_{in}} \right] \tag{3.35}$$

where the voltage reflection coefficient at the input is related to the voltage reflection coefficient at the load as

$$\hat{\Gamma}_{in} = \hat{\Gamma}_L e^{-j2\beta\mathscr{L}} \tag{3.36}$$

Normalize the input impedance by dividing it by the characteristic impedance:

$$\begin{aligned} \hat{z}_{in} &= \frac{\hat{Z}_{in}}{Z_C} \\ &= \frac{1 + \hat{\Gamma}_{in}}{1 - \hat{\Gamma}_{in}} \\ &= r + jx \end{aligned} \tag{3.37}$$

Write the reflection coefficient at the input in rectangular form as

$$\begin{aligned} \hat{\Gamma}_{in} &= \hat{\Gamma}_L e^{-j2\beta\mathscr{L}} \\ &= p + jq \end{aligned} \tag{3.38}$$

Hence

$$\begin{aligned} \hat{z}_{in} &= r + jx \\ &= \frac{1 + p + jq}{1 - p - jq} \end{aligned} \tag{3.39}$$

Solving by matching the real and imaginary parts on both sides gives two equations:

$$\left(p - \frac{r}{r+1} \right)^2 + q^2 = \frac{1}{(r+1)^2} \quad \left(\begin{array}{l} \text{Circles of radius } \dfrac{1}{r+1} \\[2mm] \text{centered at } p = \dfrac{r}{r+1} \end{array} \right) \tag{3.40a}$$

$$(p-1)^2 + \left(q - \frac{1}{x}\right)^2 = \frac{1}{x^2} \quad \left(\begin{array}{l} \text{Circles of radius } \dfrac{1}{x} \\ \text{centered at } p = 1 \text{ and } q = \dfrac{1}{x} \end{array} \right) \quad (3.40b)$$

Equation (3.40a) is plotted in Fig. 3.9(a), and (3.40b) is plotted in Fig. 3.9(b). Figure 3.10 shows a complete Smith chart.

Where is the normalized load impedance for a matched line, $\hat{Z}_L = Z_C$, plotted? Where is the normalized load impedance for a purely reactive load, $\hat{Z}_L = 1/j\omega C$ or $\hat{Z}_L = j\omega L$, plotted?

The Smith chart relates the real and imaginary parts of $\hat{z}_{in} = r + jx$ to the magnitude ($|\hat{\Gamma}_{in}| \leq 1$) and angle of the reflection coefficient *at the same point*: $\hat{\Gamma}_{in} = |\hat{\Gamma}_{in}| \angle \theta_{\hat{\Gamma}_{in}}$, as shown in Fig. 3.11. To get the input impedance at some other point on the line you simply rotate either toward the generator (TG scale) or toward the load (TL scale), as shown in Fig. 3.12.

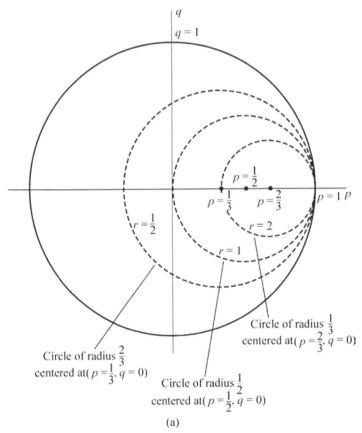

(a)

FIGURE 3.9

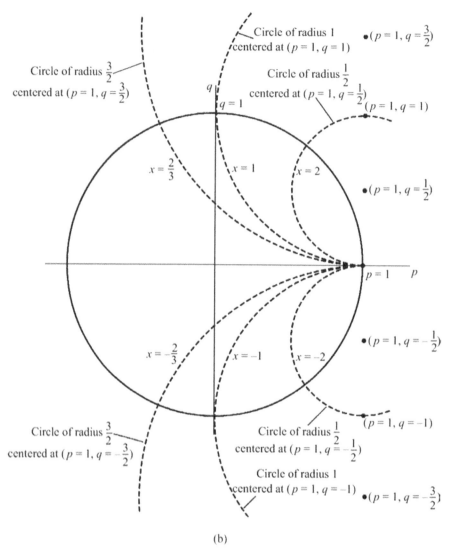

(b)

FIGURE 3.9 (*Continued*)

EXAMPLE

Consider a coaxial cable whose interior dielectric is polyethylene ($\varepsilon_r = 2.25$) and is of length 10 m. The line is driven by a source whose frequency is 34 MHz. The characteristic impedance is $Z_C = 50\,\Omega$, and the line is terminated in a load impedance of $\hat{Z}_L = (50 + j100)\,\Omega$. Determine the input

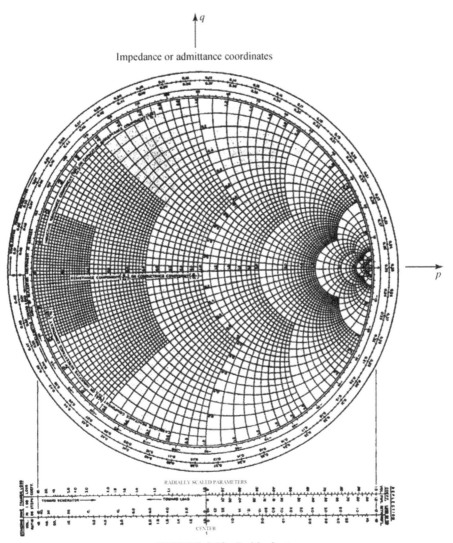

FIGURE 3.10. Smith chart.

impedance to the line. The velocity of propagation on the line is

$$v = \frac{v_0}{\sqrt{\varepsilon_r}}$$

$$= \frac{3 \times 10^8}{\sqrt{2.25}}$$

$$= 2 \times 10^8 \, \mathrm{m/s}$$

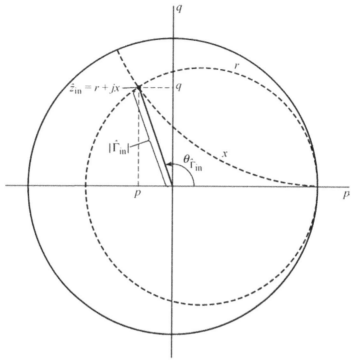

FIGURE 3.11. Relation of the normalized input impedance to the line to the voltage reflection coefficient *at that point.*

A wavelength at the operating frequency is

$$\lambda = \frac{v}{f}$$

$$= \frac{2 \times 10^8}{34 \times 10^6}$$

$$= 5.882\,\text{m}$$

Therefore, the length of the line in terms of wavelength is

$$\mathcal{L} = 10\,\text{m} \times \frac{1}{5.882\,\text{m}/\lambda}$$

$$= 1.70\,\lambda$$

The normalized load impedance is

$$\hat{z}_L = \frac{50 + j100}{50}$$

$$= 1 + j2$$

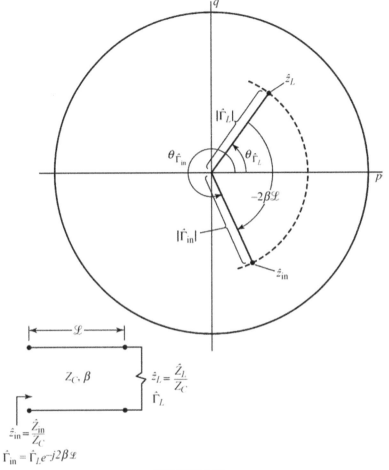

FIGURE 3.12

Plot this on the Smith chart, as shown in Fig. 3.13. Now rotate 1.7λ toward the generator (TG) to give a normalized input impedance of

$$\hat{z}_{in} = 0.29 - j0.82$$

Unnormalizing this gives the input impedance to the line as

$$\hat{Z}_{in} = \hat{z}_{in}Z_C$$
$$= (14.5 - j41)\,\Omega$$

The exact results are

$$\hat{Z}_{in} = (14.52 - j40.52)\,\Omega$$

Impedance or admittance coordinates

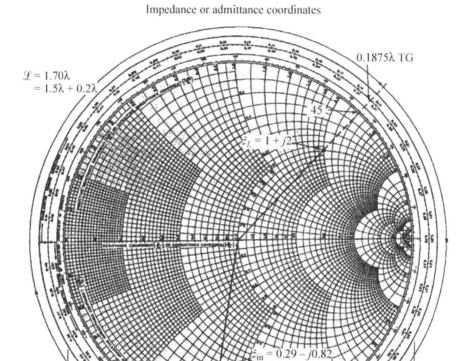

FIGURE 3.13

Observe that one full rotation around the Smith chart occurs for 0.5λ. This shows that *the input impedance replicates for lengths of line that are multiples of a half-wavelength*, as we showed previously. Hence, in rotating the full electrical length of the line, 1.7λ, we rotate three full revolutions (1.5λ) plus an additional 0.2λ.

Additionally, we can determine the magnitude of the reflection coefficient by measuring the distance from the center of the chart to the point plotted (using a compass) and transferring this length to one of the bottom scale marked REFLECTION COEFF. VOL. giving

$$\left|\hat{\Gamma}_L\right| = \left|\hat{\Gamma}_{in}\right| = 0.71$$

This shows that *the magnitude of the reflection coefficient is the same at all points on the transmission line; only the angle of the reflection coefficient varies at points along the line.*

The angle of the load reflection coefficient is read off the outer scale marked ANGLE OF REFLECTION COEFFICIENT IN DEGREES as 45°. Hence the load reflection coefficient is

$$\hat{\Gamma}_L = 0.71\angle 45°$$

The reflection coefficient at the input to the line is similarly read off at that point as

$$\hat{\Gamma}_{in} = 0.71\angle -99°$$

The VSWR can also be read off the chart by measuring (with a compass) the magnitude of the reflection coefficient and transferring that to the bottom scale marked STANDING WAVE VOL. RATIO, giving VSWR = 5.8, and the exact value is VSWR = 5.83.

Moving in the correct direction around the Smith chart in determining input impedance from the load impedance (TG) or the load impedance from the input impedance (TL) is very important.

EXAMPLE

The measured input impedance to a line is $\hat{Z}_{in} = (20 - j40)\ \Omega$ and the load impedance is $\hat{Z}_L = (20 + j40)\ \Omega$. Determine the length of the line in wavelengths if its characteristic impedance is $Z_C = 100\ \Omega$.

Plot the normalized input and load impedances on the chart shown in Fig. 3.14:

$$\hat{z}_{in} = 0.2 - j0.4$$
$$\hat{z}_L = 0.2 + j0.4$$

The length of the line in wavelengths is the circumferential distance between the two points plotted. But there are two circumferential distances between the two points; which one is correct? The simple way of determining the correct distance is simply to start at the plotted load impedance and go toward the input impedance. This requires movement TOWARD THE GENERATOR or a clockwise movement from $\hat{z}_L(0.062\lambda, \text{TG})$ to $\hat{z}_{in}(0.438\lambda, \text{TG})$, giving the line length as

$$\mathcal{L} = 0.438\lambda - 0.062\lambda$$
$$= 0.376\lambda$$

Impedance or admittance coordinates

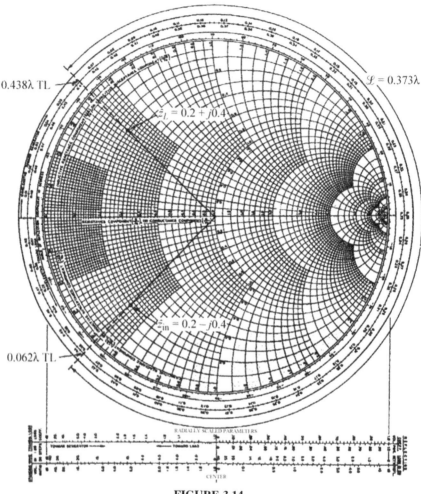

FIGURE 3.14

We could have started alternatively at the plotted input impedance and go toward the load impedance. This would require movement TOWARD THE LOAD, or counterclockwise movement from $\hat{z}_{in}(0.062\lambda, \text{TL})$ to $\hat{z}_L(0.438\lambda, \text{TL})$, giving the same result. Note that this electrical length is unique only within a multiple of a half-wavelength since the input impedance replicates for line lengths that are a multiple of a half-wavelength. So the answer could be $\mathcal{L} = 0.376\lambda + 0.5\lambda = 0.876\lambda$ or $\mathcal{L} = 0.376\lambda + 1.0\lambda = 1.376\lambda$, and so on. Knowing the velocity of propagation on the line and the frequency of the source, we can determine the physical length from

$$\mathcal{L}(\text{m}) = \mathcal{L}(\lambda) \times \left(\lambda = \frac{v}{f} \text{ m} \right)$$

For example, if the line is being operated at a frequency of 30 MHz and the velocity of propagation is 250 m/μs, the wavelength is $\lambda = 8.333$ m and the physical length is $\mathcal{L} = 3.11$ m. The VSWR can be read off as VSWR $= 5.8$.

3.11 EFFECTS OF LINE LOSSES

Two factors contribute to transmission-line losses:

1. The resistance of the line conductors
2. Losses in the surrounding dielectric medium, as in the case of the dielectric within a coaxial cable or the dielectric board of a PCB

These losses are both frequency dependent and can easily be included in the phasor (frequency-domain) transmission-line equations. Their inclusion in the time-domain transmission-line equations, however, is problematic.

The (phasor) transmission-line equations for a *lossy* line can again be derived from the per-unit-length model in Fig. 3.15, where r models the conductor losses and g models the losses in the surrounding dielectric:

$$\frac{d\hat{V}(z)}{dz} = -\underbrace{(r+j\omega l)}_{\hat{z}}\hat{I}(z)$$

$$\frac{d\hat{I}(z)}{dz} = -\underbrace{(g+j\omega c)}_{\hat{y}}\hat{V}(z)$$

(3.41)

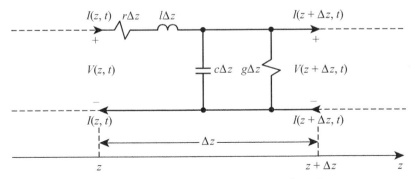

FIGURE 3.15. Per-unit-length model of a *lossy* line.

The uncoupled second-order phasor transmission-line equations are

$$
\frac{d^2\hat{V}(z)}{dz^3} - \underbrace{\hat{z}\hat{y}}_{\hat{\gamma}^2}\, \hat{V}(z) = 0
$$
$$
\frac{d^2\hat{I}(z)}{dz^3} - \underbrace{\hat{y}\hat{z}}_{\hat{\gamma}^2}\, \hat{I}(z) = 0
$$

(3.42)

The general solution is

$$
\hat{V}(z) = \hat{V}^+ e^{-\alpha z} e^{-j\beta z} + \hat{V}^- e^{\alpha z} e^{j\beta z}
$$
$$
\hat{I}(z) = \frac{\hat{V}^+}{\hat{Z}_C} e^{-\alpha z} e^{-j\beta z} - \frac{\hat{V}^-}{\hat{Z}_C} e^{\alpha z} e^{j\beta z}
$$

(3.43)

where

$$
\hat{Z}_C = \sqrt{\frac{r+j\omega l}{g+j\omega c}}
$$

(3.44)

$$
\hat{\gamma} = \sqrt{\hat{z}\hat{y}}
$$
$$
= \alpha + j\beta
$$

(3.45)

The terms $e^{\pm\alpha z}$ in the general solution in (3.43) represent *attenuation* of the amplitude due to the losses in the conductors and in the surrounding medium. Note that you can obtain all the previous results for the case of a lossy line simply by replacing $Z_C \to \hat{Z}_C$ and $j\beta \to \hat{\gamma} = \alpha + j\beta$ in the corresponding lossless line results.

Skin depth is the propensity of currents to migrate toward the outer surfaces of the conductors as their frequency is increased (Fig. 3.16), thereby increasing their resistance at a rate of $\sqrt{f}$. The skin depth is

$$
\delta = \frac{1}{\sqrt{\pi f \mu_0 \sigma}}
$$

(3.46)

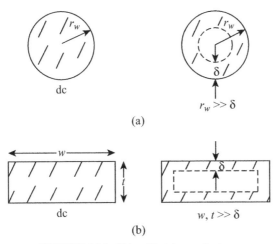

FIGURE 3.16. Skin effect in conductors.

In the case of wires, the resistance has two frequency regions: the dc region and the high-frequency region:

$$r_{dc} = \frac{1}{\sigma \pi r_w^2} \quad \Omega/m \qquad r_w < 2\delta \tag{3.47a}$$

$$r_{hf} \cong \frac{1}{\sigma 2 \pi r_w \delta} \quad \Omega/m \qquad r_w > 2\delta \tag{3.47b}$$

In the case of lands on PCBs, the situation is more complicated but can similarly be approximated by two frequency regions:

$$r_{dc} = \frac{1}{\sigma w t} \quad \Omega/m \tag{3.48a}$$

$$r_{hf} \cong \frac{1}{2\sigma \delta (w + t)} \quad \Omega/m \qquad \max(w, t) > 2\delta \tag{3.48b}$$

Figure 3.17 illustrates this frequency dependence of the resistance of the conductors.

Figure 3.18 illustrates the point that bound dipoles in dielectrics resist instantaneous rotation, thereby creating dielectric losses. At higher frequencies, the electric field changes direction more rapidly. The bound dipoles in the dielectric are more resistant to change direction to align with this.

To account for dielectric losses, we employ a complex relative permittivity:

$$\varepsilon_r = \varepsilon_r' - j\varepsilon_r'' \tag{3.49}$$

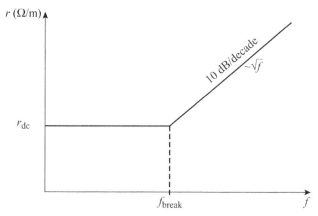

FIGURE 3.17. Skin effect losses of conductors.

The imaginary part, ε_r'', accounts for losses in the dielectric. The dielectric losses are represented in terms of a *loss tangent* that is tabulated in various handbooks at various frequencies as the ratio of the imaginary and real parts of the complex relative permittivity:

$$\tan\theta = \frac{\varepsilon_r''}{\varepsilon_r'} \tag{3.50}$$

The per-unit-length conductance, g, in the per-unit-length equivalent circuit in Fig. 3.15 is given in terms of this loss tangent, and for a *homogeneous medium* is given in terms of the per-unit-length capacitance c and radian frequency of the source as

$$\boxed{g = \omega c \tan\theta} \tag{3.51}$$

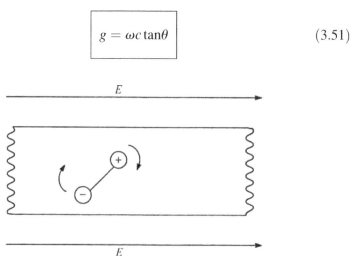

FIGURE 3.18. Resistance of bound dipoles in dielectrics to instantaneous rotation, thereby creating losses.

Observe that (3.51) appears to suggest that the per-unit-length conductance g appears to increase linearly with frequency. However, this is not the case, since the loss tangent has a frequency dependence in the form of resonances over certain frequency ranges. For example, glass-epoxy (FR4), commonly used to construct PCBs, has a loss tangent of $\tan\theta \cong 0.02$ over the frequency range $1\,\text{MHz} < f < 100\,\text{MHz}$ and falls off either side of this frequency range. From (3.51) we see that since $\tan\theta = 0.02$, $g = 0.02\omega c$ and the parallel admittance g is $\frac{1}{50}$ the magnitude of the parallel admittance ωc in the per-unit-length equivalent circuit of Fig. 3.15.

The *characteristic impedance* of a lossy line is, in general, complex:

$$\hat{Z}_C = \sqrt{\frac{r+j\omega l}{g+j\omega c}} = Z_C \angle \theta_{Z_C} \tag{3.52}$$

The velocity of propagation is properly obtained from the imaginary part of the propagation constant, β:

$$\begin{aligned} \hat{\gamma} &= \sqrt{(r+j\omega l)(g+j\omega c)} \\ &= \alpha + j\beta \end{aligned} \tag{3.53}$$

as

$$v = \frac{\omega}{\beta} \tag{3.54}$$

The losses of a practical transmission line are expected to be small; otherwise, the transmission line would be useless for transmission of signals. Hence it is practical to make *small-loss approximations* to determine these important parameters. These *small-loss approximations* are obtained for the frequency regions

$$\begin{aligned} r &\ll \omega l \\ g &\ll \omega c \end{aligned} \tag{3.55}$$

Substituting these assumptions into the appropriate equations and making approximations gives

$$\hat{Z}_C = \sqrt{\frac{r+j\omega l}{g+j\omega c}}$$

$$\cong \sqrt{\frac{j\omega l}{j\omega c}} \qquad \begin{cases} r \ll \omega l \\ g \ll \omega c \end{cases}$$

$$= \sqrt{\frac{l}{c}} = Z_C$$

$$\hat{\gamma} = \sqrt{(r+j\omega l)(g+j\omega c)}$$

$$= \sqrt{j\omega l \, j\omega c \left(1 + \frac{r}{j\omega l}\right)\left(1 + \frac{g}{j\omega c}\right)}$$

$$\cong j\omega\sqrt{lc}\sqrt{1 - j\left(\frac{r}{\omega l} + \frac{g}{\omega c}\right)} \qquad \begin{cases} r \ll \omega l \\ g \ll \omega c \end{cases} \qquad (3.57)$$

$$\cong j\omega\sqrt{lc}\left[1 - j\frac{1}{2}\left(\frac{r}{\omega l} + \frac{g}{\omega c}\right)\right]$$

$$= \underbrace{\frac{1}{2}\left(\frac{r}{Z_C} + gZ_C\right)}_{\alpha} + j\underbrace{\omega\sqrt{lc}}_{\beta}$$

and using the binomial theorem,

$$\sqrt{1 \pm x} \cong 1 \pm \frac{1}{2}x \qquad x \ll 1$$

So for small losses, $r \ll \omega l, g \ll \omega c$:

$$\boxed{\hat{Z}_C \cong \sqrt{\frac{l}{c}}} \qquad (3.58)$$

$$\boxed{v = \frac{\omega}{\beta} \cong \frac{1}{\sqrt{lc}}} \qquad (3.59)$$

$$\boxed{\alpha \cong \frac{1}{2}\left(\frac{r}{Z_C} + gZ_C\right)} \tag{3.60}$$

$$\boxed{\beta \cong \omega\sqrt{l\,c}} \tag{3.61}$$

EXAMPLE

Consider a stripline with $s = 20$ mils, $w = 5$ mils, and $t = 1.4$ mils $(1 - \text{oz copper})$. For a FR-4 PCB inserted between the two ground planes, $\varepsilon_r \cong 4.7$ and $\tan\theta \cong 0.02$, 1 to 100 MHz:

$$l = 0.461\,\mu\text{H/m}$$
$$c = 113.2\,\text{pF/m}$$
$$g = 14.2 \times 10^{-12}f \qquad \text{S/m}$$
$$r = r_{\text{dc}} = 3.818\,\Omega/\text{m} \qquad f < 23\,\text{MHz}$$
$$r = r_{\text{hf}} = 8.05 \times 10^{-4}\sqrt{f}\,\Omega/\text{m} \qquad f > 23\,\text{MHz}$$

The low-loss approximations ($r < \omega l, g < \omega c$) are valid for $f > 1.32$ MHz and yield

$$Z_C \cong \sqrt{\frac{l}{c}} = 63.8\,\Omega$$

$$v \cong \frac{1}{\sqrt{lc}} = 1.384 \times 10^8\,\text{m/s}$$
$$= \frac{v_0}{\sqrt{\varepsilon_r}}$$

$$\alpha \cong \frac{1}{2}\left(\frac{r}{Z_C} + gZ_C\right) \cong 6.307 \times 10^{-6}\sqrt{f} + 4.53 \times 10^{-10}f$$

Figure 3.19 gives a plot of the exact result (no approximations) for the magnitude and angle of the complex characteristic impedance for this problem:

$$\hat{Z}_C = \sqrt{\frac{r + j\omega l}{g + j\omega c}} = Z_C\angle\theta_{Z_C}$$

Observe that the magnitude is the real value of 63.8 Ω above about 1.32 MHz.

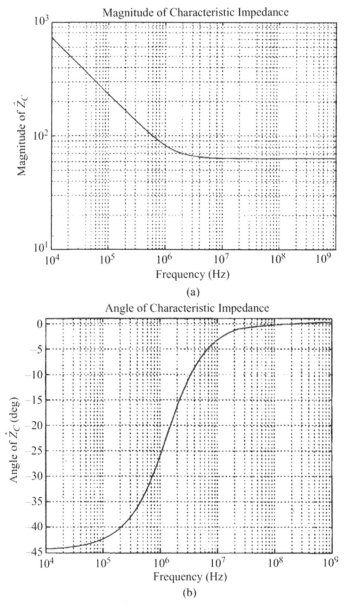

FIGURE 3.19. Plot of (a) the magnitude and (b) the angle (exact, no approximations) for the characteristic impedance for the stripline.

Figure 3.20(a) gives the exact value for the attenuation constant for this stripline problem, while Fig. 3.20(b) shows that the velocity of propagation is the approximate value for a lossless line of around 1.384×10^8 m/s above about 1.32 MHz. Figure 3.20 computes the exact value of the complex

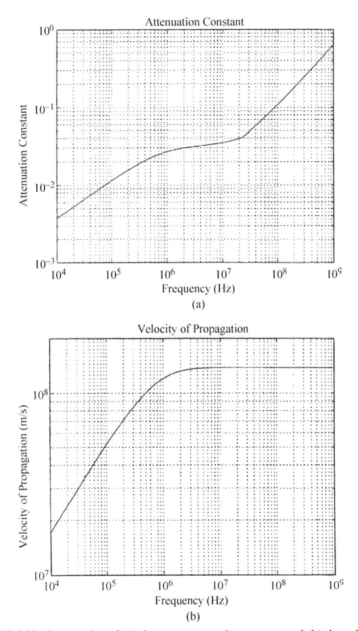

FIGURE 3.20. Computation of (a) the exact attenuation constant and (b) the velocity of propagation (no approximations).

propagation constant,

$$\hat{\gamma} = \sqrt{(r+j\omega l)(g+j\omega c)}$$
$$= \alpha + j\beta$$

but the velocity of propagation is computed from $v = \omega/\beta$. The figure shows that computations assuming lossless lines in fact give a reasonably accurate picture of how an actual line behaves. In the near future, losses may begin to significantly affect system behavior. When we consider the fact that the fundamental frequencies of digital clock and data signals today exceed 100 MHz, low-loss approximations are in fact valid.

EXAMPLE

As a final example we compute the response of a stripline with and without losses whose cross section is shown in Fig. 3.21(a) and whose terminations are shown in Fig. 3.21(b). It is driven by a 5-V 50-MHz clock waveform having 500-ps rise and fall times and a 50% duty cycle, shown in Fig. 3.21(c).

For a *lossless line*, the load voltage after three cycles of $V_S(t)$ is computed with the following PSPICE program and is shown in Fig. 3.22.

The PSPICE program is

```
EXAMPLE
VS 1 0 PULSE(0 5 0 500P 500P 9.5N 20N)
RS 1 2 10
T 2 0 3 0 Z0=63.816 TD=0.9174N
RL 3 0 1E8
.TRAN 10P 60N 0 10P
.PRINT TRAN V(3)
.PROBE
.END
```

Notice that the response to the initial pulse $0 < t < 10.5$ ns has not gotten into steady state when the pulse turns off and remains at zero for 10.5 ns $< t < 20$ ns. But reflections due to that initial half of the cycle continue for 10.5 ns $< t < 20$ ns even though the pulse waveform is turned off. After the first cycle of $V_S(t)$, the load voltage essentially reaches steady state.

The lossless and lossy line responses are compared over the third cycle of $V_S(t)$, 40 ns $< t < 60$ ns, in Fig. 3.23. Note that the losses do not degrade the

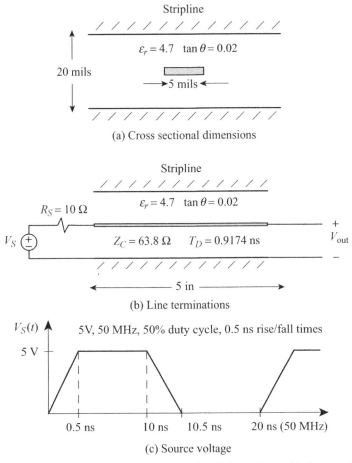

(a) Cross sectional dimensions

(b) Line terminations

(c) Source voltage

FIGURE 3.21. Stripline problem: (a) cross-sectional dimensions; (b) line terminations;. (c) source voltage.

waveform substantially for this terminal voltage response of $V_L(t)$ after it is in steady state [essentially after the first half cycle of $V_S(t)$].

The *lossy line* response is computed using superposition and adding the time-domain responses of the dc component of $V_S(t)$ and the first 19 sinusoidal harmonic components: dc and 50, 150, 250, 350, 450, 550, 650, 750, 850, and 950 MHz, as illustrated in Fig. 3.24 by embedding the transmission line and the termination impedances in a linear system with input $V_S(t)$ and output $V_L(t)$, as shown in Fig. 3.25. As discussed in Part I, this technique is valid only for the time interval when the response has reached steady state.

The frequency-domain transfer function is computed at each of the harmonics using the phasor solution from Fig. 3.26. The frequency-domain

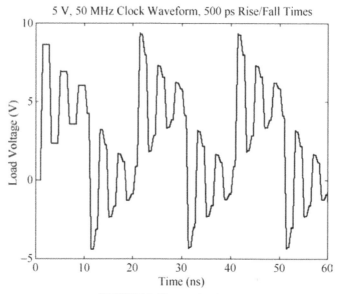

FIGURE 3.22. Load voltage.

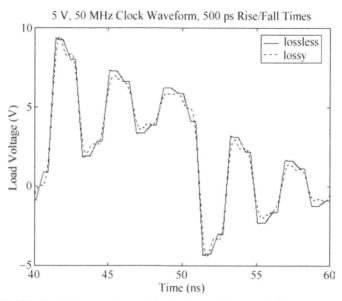

FIGURE 3.23. Comparison of the lossless and lossy stripline reponses.

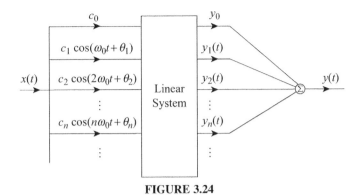

FIGURE 3.24

transfer function at each of the harmonic frequencies is computed from (3.32):

$$\hat{V}(z) = \frac{1 + \hat{\Gamma}_L e^{-2\alpha(\mathcal{L}-z)} e^{-j2\beta(\mathcal{L}-z)}}{1 - \hat{\Gamma}_S \hat{\Gamma}_L e^{-2\alpha\mathcal{L}} e^{-j2\beta\mathcal{L}}} \frac{Z_C}{\hat{Z}_S + Z_C} \hat{V}_S e^{-\alpha z} e^{-j\beta z} \tag{3.32}$$

as

$$\hat{H}(j\omega) = \frac{\hat{V}(\mathcal{L})}{\hat{V}_S}$$

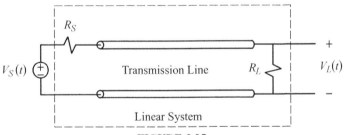

FIGURE 3.25

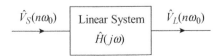

FIGURE 3.26. Computing the phasor transfer function.

From our previous computations, the low-loss approximations are valid for this stripline for frequencies at and above the fundamental frequency of 50 MHz:

$$r = r_{hf} = 8.05 \times 10^{-4}\sqrt{f} \quad \Omega/m \qquad f > 23\,\text{MHz}$$

$$g = 14.2 \times 10^{-12}f \qquad \text{S/m}$$

$$\alpha \cong \frac{1}{2}\left(\frac{r}{Z_C} + gZ_C\right)$$

$$v = \frac{\omega}{\beta} \cong 1.384 \times 10^{8}\,\text{m/s}$$

$$Z_C \cong \sqrt{\frac{l}{c}} = 63.8\,\Omega$$

For this problem

$$\hat{\Gamma}_S = \frac{R_S - Z_C}{R_S + Z_C}$$

$$= \frac{10 - 63.8}{10 + 63.8}$$

$$= -0.7291$$

$$\hat{\Gamma}_L = \frac{R_L - Z_C}{R_L + Z_C}$$

$$= \frac{\infty - 63.8}{\infty + 63.8}$$

$$= +1$$

so that

$$\hat{H}(j\omega) = \frac{\hat{V}(\mathscr{L})}{\hat{V}_S}$$

$$= \frac{2}{1 + 0.7291e^{-2\alpha\mathscr{L}}e^{-j2\beta\mathscr{L}}}0.865e^{-\alpha\mathscr{L}}e^{-j\beta\mathscr{L}}$$

The magnitudes and angles of the first 19 harmonics of $V_S(t)$ (omitting the even harmonics whose magnitudes are zero because of the 50% duty cycle of the pulse train) are given in Table 3.1 along with the magnitudes and

TABLE 3.1. Harmonic Components of $V_S(t)$ and the Transfer Function for the Stripline Problem in Fig. 3.21

Frequency (MHz)	$|c_n|$	$\angle c_n$	$|\hat{H}(j\omega)|$	$\angle\hat{H}(j\omega)$
Dc	2.5	0	1	0
50	3.1798	−94.5	1.041	−2.8
150	1.0512	−103.5	1.506	−11.57
250	0.6204	−112.5	4.471	−54.51
350	0.4322	−121.5	2.138	−158.3
450	0.3250	−130.5	1.155	−173.1
550	0.2547	−139.5	0.9918	179.7
650	0.2045	−148.5	1.191	171.6
750	0.1664	−157.5	2.297	151.8
850	0.1364	−166.5	3.402	50.03
950	0.1119	−175.5	1.384	13.67

angles of the transfer function. Hence the load voltage including losses is computed as

$$V_L(t) = 2.5 + \sum_{n=1}^{19} |c_n| |\hat{H}(j\omega)| \cos\left(n\omega_0 t + \angle c_n + \angle\hat{H}(j\omega)\right)$$

where $\omega_0 = 2\pi f_0$ and $f_0 = 50\,\text{MHz}$.

3.12 LUMPED-CIRCUIT APPROXIMATIONS FOR ELECTRICALLY SHORT LINES

If the line is electrically short at *the highest significant frequency of the waveform* (its bandwidth), we can approximately model it with lumped-circuit models shown in Fig. 3.27 and avoid solving the transmission-line equations. The advantage here is that although PSPICE contains an exact model for a lossless line, we can incorporate line losses into this lumped-circuit model. However, the disadvantage is that frequency-dependent losses (as they invariably are) cannot be incorporated since the lumped-circuit elements must be frequency independent.

EXAMPLE

For the stripline example shown in Fig. 3.21, we can model the *lossless* line with approximate lumped-pi models as shown in Fig. 3.28.

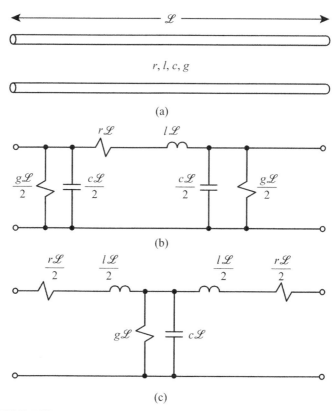

FIGURE 3.27. Lumped-circuit approximate models of electrically short lines.

The PSPICE program is

```
EXAMPLE
VS 1 0 PULSE(0 5 0 500P 500P 9.5N 20N)
*VS 1 0 AC 1 0
* TRANSMISSION LINE MODEL
RS1 1 2 10
T 2 0 3 0 Z0=63.816 TD=0.9174N
RL1 3 0 1E8
* ONE-PI SECTION
RS2 1 4 10
C11 4 0 7.188P
L11 4 5 58.55N
C12 5 0 7.188P
RL2 5 0 1E8
* TWO-PI SECTIONS
RS3 1 6 10
```

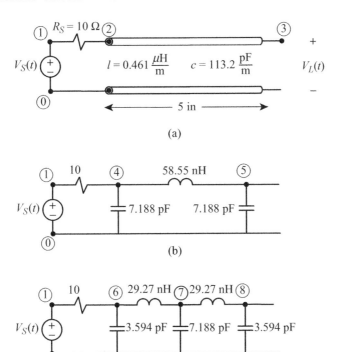

FIGURE 3.28. Lumped-circuit approximate model of the stripline example in Fig. 3.21.

```
C21 6 0 3.594P
L21 6 7 29.27N
C22 7 0 7.188P
L22 7 8 29.27N
C23 8 0 3.594P
RL3 8 0 1E8
.TRAN 10P 20N 0 10P
.PRINT TRAN V(3) V(5) V(8)
.PROBE
*.AC DEC 100 1E7 1E9
*.PRINT AC VM(3) VM(5) VM(8)
.END
```

Using PSPICE to solve these gives the time-domain solution shown in Fig. 3.29. The line is one wavelength long at $1/T_D = 1.09$ GHz and is electrically short at 109 MHz. Hence the line is electrically short only at the fundamental frequency of the pulse train of 50 MHz. Hence it is not surprising that these lumped-circuit models do a poor job of representing

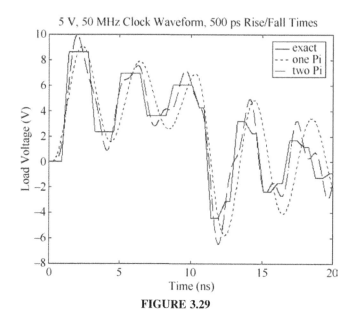

FIGURE 3.29

the line response since the bandwidth of the waveform is $BW = 1/\tau_r = 2\,GHz$.

If we slow the rise and fall times to 2 ns, the bandwidth of the waveform is reduced to $BW = 1/\tau_r = 500\,MHz$. Figure 3.30 shows the time-domain response comparing the exact transmission-line model with one-pi and two-pi section models. The two-pi model gives a good correlation, but the one-pi model is considerably in error.

If we slow the rise and fall times further to 5 ns, the bandwidth of the waveform is reduced to $BW = 1/\tau_r = 200\,MHz$. Figure 3.31 shows the time-domain response comparing the exact transmission-line model with one-pi and two-pi models. The one-pi model provides better predictions, and the two-pi model gives a closer correlation.

We can easily show the frequency-domain transfer function of the line by changing a few lines of the PSPICE program (those noted with a star as comment lines). The magnitude of the frequency response is computed as shown in Fig. 3.32. Observe that compared to the exact transmission-line model, the one-pi section model gives adequate predictions up to about 200 MHz and the two-pi model gives adequate predictions up to about 500 MHz. This explains why the response to the 2-ns rise and fall time pulse, which has a bandwidth of 500 MHz, is predicted adequately only by a two-pi model, but the response to the 5-ns rise and fall time pulse, which has a bandwidth of 200 MHz, is adequately predicted by both the two-pi and one-pi models.

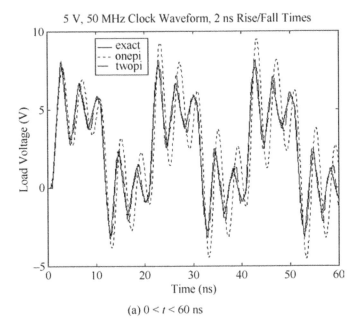

(a) $0 < t < 60$ ns

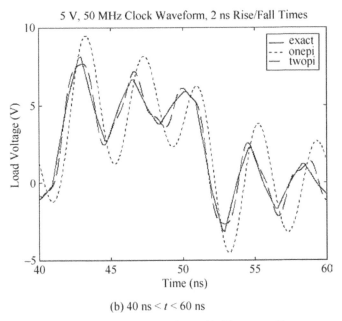

(b) 40 ns $< t < 60$ ns

FIGURE 3.30. (a) $0 < t < 60$ ns; (b) 40 ns $< t < 60$ ns.

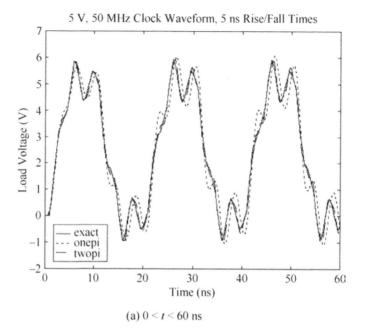

(a) $0 < t < 60$ ns

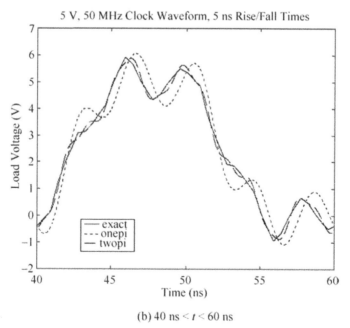

(b) 40 ns $< t < 60$ ns

FIGURE 3.31. (a) $0 < t < 60$ ns; (b) 40 ns $< t < 60$ ns.

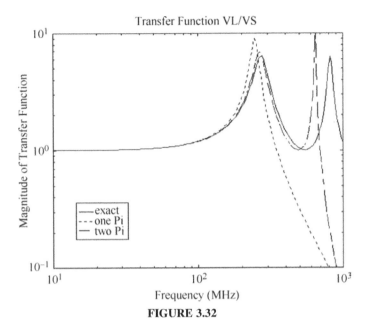

FIGURE 3.32

Figure 3.32 shows that using more than one lumped-pi section to try to extend the frequency range doesn't gain very much and results in a larger and larger lumped-circuit model for PSPICE to solve. So if the line is electrically short, a single-section lumped-pi model works fine. If the line is *not* electrically short, you have no other choice but to solve the transmission-line equations.

3.13 CONSTRUCTION OF MICROWAVE CIRCUIT COMPONENTS USING TRANSMISSION LINES

The inductance and capacitance of the leads attaching a discrete resistor, capacitor, or inductor to the circuit can cause significant deterioration of its performance at microwave frequencies. Shorted transmission lines on PCBs are used to simulate lumped-circuit elements in microwave circuits.

The input impedance to a transmission line in its most general form is

$$\hat{Z}_{in} = Z_C \left[\frac{1 + \hat{\Gamma}_L e^{-j2\beta\mathscr{L}}}{1 - \hat{\Gamma}_L e^{-j2\beta\mathscr{L}}} \right]$$

Consider the case of a line that is a quarter-wavelength long, $\mathscr{L} = \lambda/4$, shown in Fig. 3.33.

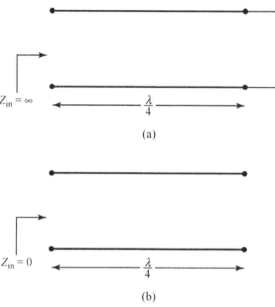

FIGURE 3.33

In the case of a short-circuit load,

$$\hat{\Gamma}_L = \frac{0 - Z_C}{0 + Z_C}$$

$$= -1$$

$$e^{-j2\beta \mathscr{L}} = e^{-j4\pi(\mathscr{L}/\lambda)}$$

$$e^{-j2\beta \mathscr{L}} = e^{-j\pi}$$

$$= -1 \qquad \mathscr{L} = \frac{1}{4}\lambda$$

$$\hat{\Gamma}_{in} = \hat{\Gamma}_L e^{-j2\beta \mathscr{L}}$$

$$= (-1)(-1)$$

$$= 1$$

$$\hat{Z}_{in} = Z_C \frac{1+1}{1-1}$$

$$= \infty \begin{cases} \hat{Z}_L = \text{short circuit} \\ \mathscr{L} = \dfrac{\lambda}{4} \end{cases}$$

Hence a quarter-wavelength line having a short-circuit load appears as an open circuit!

In the case of an open-circuit load, $\Gamma_L = +1$:

$$\hat{Z}_{in} = Z_C \frac{1-1}{1+1}$$

$$= 0 \begin{cases} \hat{Z}_L = \text{open circuit} \\ \mathcal{L} = \dfrac{\lambda}{4} \end{cases}$$

Hence a quarter-wavelength line having an open-circuit load appears as a short circuit!

Can we construct other circuit elements with a short-circuited section of transmission line of different lengths?

$$\hat{Z}_{in} = Z_C \left[\frac{1 - e^{-j2\beta\mathcal{L}}}{1 + e^{-j2\beta\mathcal{L}}} \right] \begin{cases} \text{short-circuit load} \\ \hat{Z}_L = 0 \end{cases}$$

But this can be written as

$$\hat{Z}_{in} = Z_C \left[\frac{1 - e^{-j2\beta\mathcal{L}}}{1 + e^{-j2\beta\mathcal{L}}} \right]$$

$$= Z_C \frac{e^{-j\beta\mathcal{L}}}{e^{-j\beta\mathcal{L}}} \left[\frac{e^{j\beta\mathcal{L}} - e^{-j\beta\mathcal{L}}}{e^{j\beta\mathcal{L}} + e^{-j\beta\mathcal{L}}} \right]$$

$$= jZ_C \tan\left(2\pi \frac{\mathcal{L}}{\lambda}\right) \begin{cases} \text{short-circuit load} \\ \hat{Z}_L = 0 \end{cases}$$

which can be plotted as shown in Fig. 3.34.

We can construct a 10-pF capacitor at 1 GHz using a short-circuited $Z_C = 50\,\Omega$ line. The impedance of a 10-pF lumped-circuit capacitor at 1 GHz is

$$-j\frac{1}{\omega C} = -j15.92\,\Omega$$

Compare this to the equation for input impedance to a short-circuited transmission line:

$$jZ_C \tan\left(2\pi \frac{\mathcal{L}}{\lambda}\right) = -j15.92$$

Solve to give

$$\frac{\mathcal{L}}{\lambda} = -0.049$$

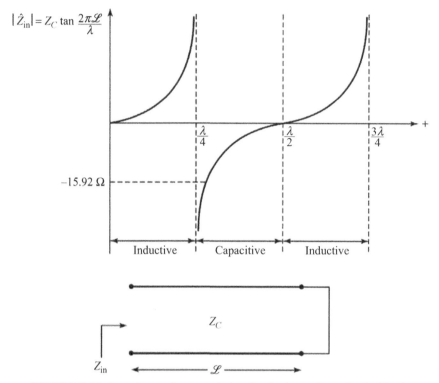

FIGURE 3.34. Impedance of a transmission line having a short-circuited load.

or

$$\frac{\mathscr{L}}{\lambda} = -0.049 + 0.5 = 0.451$$

If the line is air filled, $v = 3 \times 10^8$ m/s, and the physical length (at 1 GHz) is

$$\mathscr{L} = 0.451 \times 0.3$$
$$= 13.53\,\text{cm}$$

PROBLEMS

3.1 Verify by direct substitution that the solutions for the phasor voltage and current given in (3.6) satisfy the phasor transmission-line equations given in (3.4).

3.2 Verify by direct substitution that the time-domain solutions for the line voltage and current given in (3.8) satisfy the time-domain transmission-line equations given in (2.4).

3.3 For the phasor transmission line shown in Fig. 3.1, $f = 5$ MHz, $v = 3 \times 10^8$ m/s, $\mathcal{L} = 78$ m, $Z_C = 50\,\Omega$, $V_S = 50\angle 0°$, $Z_S = 20 - j30\,\Omega$, $\hat{Z}_L = 200 + j500\,\Omega$. Determine (**a**) the line length as a fraction of a wavelength, (**b**) the voltage reflection coefficient at the load and at the input to the line, (**c**) the input impedance to the line, (**d**) the time-domain voltages at the input to the line and at the load, (**e**) the average power delivered to the load, and (**f**) the VSWR. [(a) 1.3; (b) $0.9338\angle 9.866°$, $0.9338\angle 153.9°$; (c) $11.73\angle 81.16°$; (d) $20.55\cos(10\pi \times 10^6 t + 121.3°)$, $89.6\cos(10\pi \times 10^6 t - 50.45°)$; (e) 2.77 W; (f) 29.21]

3.4 For the phasor transmission line shown in Fig. 3.1, $f = 200$ MHz, $v = 3 \times 10^8$ m/s, $\mathcal{L} = 2.1$ m, $Z_C = 100\,\Omega$, $\hat{V}_S = 10\angle 60°$, $\hat{Z}_S = 50\,\Omega$, $\hat{Z}_L = 10 - j50\,\Omega$. Determine (**a**) the line length as a fraction of a wavelength, (**b**) the voltage reflection coefficient at the load and at the input to the line, (**c**) the input impedance to the line, (**d**) the time-domain voltages at the input to the line and at the load, (**e**) the average power delivered to the load, and (**f**) the VSWR. [(a) 1.4; (b) $0.8521\angle -126.5°$, $0.8521\angle -54.5°$; (c) $192\angle -78.83°$; (d) $9.25\cos(4\pi \times 10^8 t + 46.33°)$, $4.738\cos(4\pi \times 10^8 t - 127°)$; (e) 43 mW; (f) 12.52]

3.5 For the phasor transmission line shown in Fig. 3.1, $f = 1$ GHz, $v = 1.7 \times 10^8$ m/s, $\mathcal{L} = 11.9$ cm, $Z_C = 100\,\Omega$, $\hat{V}_S = 5\angle 0°$, $\hat{Z}_S = 20\,\Omega$, $\hat{Z}_L = -j160\,\Omega$. Determine (**a**) the line length as a fraction of a wavelength, (**b**) the voltage reflection coefficient at the load and at the input to the line, (**c**) the input impedance to the line, (**d**) the time-domain voltages at the input to the line and at the load, (**e**) the average power delivered to the load, and (**f**) the VSWR. [(a) 0.7; (b) $1\angle -64.1°$, $1\angle 152°$; (c) $24.94\angle 90°$; (d) $3.901\cos(2\pi \times 10^9 t + 38.72°)$, $13.67\cos(2\pi \times 10^9 t + 38.72°)$; (e) 0 W, (f) ∞]

3.6 For the phasor transmission line shown in Fig. 3.1, $f = 600$ MHz, $v = 2 \times 10^8$ m/s, $\mathcal{L} = 53$ cm, $Z_C = 75\,\Omega$, $\hat{V}_S = 20\angle 40°$, $\hat{Z}_S = 30\,\Omega$, $\hat{Z}_L = 100 - j300\,\Omega$. Determine (**a**) the line length as a fraction of a wavelength, (**b**) the voltage reflection coefficient at the input to the line and at the load, (**c**) the input impedance to the line, (**d**) the time-domain voltages at the the input to the line and at the load, (**e**) the average power delivered to the load, and (**f**) the VSWR. [(a) 1.59; (b) $0.8668\angle -25.49°$, $0.8668\angle -90.29°$; (c) $74.62\angle -81.84°$; (d) $17.71\cos(12\pi \times 10^8 t + 19.37°)$, $24.43\cos(12\pi \times 10^6 t - 163.8°)$; (e) 0.298 W; (f) 14.02]

3.7 For the phasor transmission line shown in Fig. 3.1, $f = 1$ MHz, $v = 3 \times 10^8$ m/s, $\mathcal{L} = 108$ m, $Z_C = 300\,\Omega$, $\hat{V}_S = 100\angle 0°$, $\hat{Z}_S = 50 + j50\,\Omega$, $\hat{Z}_L = 100 - j100\,\Omega$. Determine (**a**) the line length as a fraction of a wavelength, (**b**) the voltage reflection coefficient at the load and at the input to the line, (c) the input impedance to the line, (d) the time-domain voltages at the input to the line and at the load, (e) the average power delivered to the load, and (f) the VSWR. [(a) 0.36, (b) 0.5423∠ − 139.4°, 0.5423Đ-38.6°; (c) 657.1∠ − 43.79°; (d) 99.2cos(2π × 10^{6}t − 6.127°), 46.5cos(2π× 10^{6}t − 153.3°); (e) 5.41 W; (f) 3.37]

3.8 Confirm the input and load voltages for Problem 3.3 using SPICE.

3.9 Confirm the input and load voltages for Problem 3.4 using SPICE.

3.10 Confirm the input and load voltages for Problem 3.5 using SPICE.

3.11 Confirm the input and load voltages for Problem 3.6 using SPICE.

3.12 Confirm the input and load voltages for Problem 3.7 using SPICE.

3.13 A half-wavelength dipole antenna is connected to a 100-MHz source with a 3.6-m length of 300-Ω transmission line (twin lead, $v = 2.6 \times 10^8$ m/s). The source is represented by an open-circuit voltage of 10 V and source impedance of 50 Ω, whereas the input to the dipole antenna is represented by a 73-Ω resistance in series with an inductive reactance of 42.5 Ω. The average power dissipated in the 73-Ω resistance is equal to the power radiated into space by the antenna. Determine the average power radiated by the antenna with and without the transmission line, and the VSWR on the line. [91.14 mW, 215.53 mW, 4.2]

3.14 Two identical half-wavelength dipole antennas whose input impedances 73 + j42.5 Ω are connected in parallel with half-wave transmission lines and fed from one 300-MHz 10-V, 50-Ω source by a 1.5λ-length transmission line. Determine the average power delivered to each antenna. [115 mW]

3.15 Determine an expression for the input impedance to (a) a transmission line having an open-circuit load, and (b) a transmission line having a short-circuit load. [(a) $-jZ_C(1/\tan\beta\, L)$,(b) $jZ_C\tan\beta\, L$]

3.16 Obtain an expression for the input impedance to a quarter-wavelength transmission line. If the line has an open-circuit load, what is its input impedance? If the line has a short-circuit load, what is its input impedance? [$\hat{Z}_{\text{in}} = \hat{Z}_C^2/\hat{Z}_L$, short circuit, open circuit].

3.17 Determine, using the Smith chart, the input impedance, input reflection coefficient, load reflection coefficient, and VSWR for the line of Problem 3.3. [$\hat{Z}_{\text{in}} = 2 + j12$, $\hat{\Gamma}(0) = 0.92∠154°$, $\hat{\Gamma}_L = 0.92∠7°$, VSWR = 30]

3.18 Determine, using the Smith chart, the input impedance, input reflection coefficient, load reflection coefficient, and VSWR for the line of

Problem 3.4. $[\hat{Z}_{\text{in}} = 38 - j188, \ \hat{\Gamma}(0) = 0.842\angle - 55°, \ \hat{\Gamma}_L = 0.842$
$\angle - 126°, \text{VSWR} = 12]$

3.19 Determine, using the Smith chart, the input impedance, input reflection coefficient, load reflection coefficient, and VSWR for the line of Problem 3.5. $[\hat{Z}_{\text{in}} = j25, \hat{\Gamma}(0) = 1\angle 152°, \hat{\Gamma}_L = 1\angle - 64°, \text{VSWR} = \infty]$

3.20 Determine, using the Smith chart, the input impedance, input reflection coefficient, load reflection coefficient, and VSWR for the line of Problem 3.6. [(a) 1.59; (b) $0.8668\angle - 25.49°, 0.8668\angle - 90.29°$; (c) $74.62\angle - 81.84°$; (d) $17.71\cos(12\pi \times 10^8 t + 19.37°)$, $24.43\cos(12\pi \times 10^6 t - 163.8°)$; (e) 0.298 W; (f) 14.02]

3.21 Determine, using the Smith chart, the input impedance, input reflection coefficient, load reflection coefficient, and VSWR for the line of Problem 3.7. $[\hat{Z}_{\text{in}} = 474 - j450, \ \hat{\Gamma}(0) = 0.54\angle - 39°,$
$\hat{\Gamma}_L = 0.54\angle - 139°, \text{VSWR} = 3.4]$

3.22 Determine the load impedance, VSWR, and load reflection coefficient for the following transmission lines: **(a)** $\hat{Z}_{\text{in}} = (30 - j100)\Omega$, $Z_C = 50\,\Omega$, $\mathcal{L} = 0.4\lambda$; **(b)** $\hat{Z}_{\text{in}} = 50\,\Omega$, $Z_C = 75\,\Omega$, $\mathcal{L} = 1.3\lambda$; **(c)** $\hat{Z}_{\text{in}} = (150 + j230)\Omega$, $Z_C = 100\,\Omega$, $\mathcal{L} = 0.6\lambda$; **(d)** $\hat{Z}_{\text{in}} = j250\,\Omega$, $Z_C = 100\,\Omega$, $\mathcal{L} = 0.8\lambda$.

3.23 Determine the shortest lengths of the following transmission lines as well as the VSWR and the load reflection coefficient: **(a)** $\hat{Z}_{\text{in}} = -j20\,\Omega$, $\hat{Z}_L = j50\,\Omega$, $Z_C = 100\,\Omega$; **(b)** $\hat{Z}_{\text{in}} = (50 - j200)\Omega$, $\hat{Z}_L = (12 - j50)\Omega$, $Z_C = 100\,\Omega$; **(c)** $\hat{Z}_{\text{in}} = (30 + j50)\Omega$, $\hat{Z}_L = (200 + j200)\Omega$, $Z_C = 100\,\Omega$; **(d)** $\hat{Z}_{\text{in}} = (135 + j0)\Omega$, $\hat{Z}_L = (60 - j37.5)\Omega$, $Z_C = 75\,\Omega$. [(a) 0.396λ, ∞, $1\angle 127°$; (b) 0.395λ, 10, $0.83\angle - 126°$; (c) 0.37λ, 4.2, $0.62\angle 29.5°$; (d) 0.366λ, 1.8, $0.285\angle - 96°$]

3.24 Using the Smith chart, show the following properties of lossless transmission lines: **(a)** $|\hat{\Gamma}_L| \leq 1$; **(b)** the input impedance replicates for distances separated by a multiple of a half-wavelength; **(c)** the VSWR for a line having a purely reactive load is infinite, and the magnitude of the reflection coefficient is unity; **(d)** the input impedance at any point on a line having a purely reactive load cannot have a real part; **(e)** adjacent points on a line where the input impedance is purely resistive are separated by a quarter-wavelength.

3.25 We want to determine the value of an unknown impedance, $\hat{Z}_L$, attached to a length of transmission line having a characteristic impedance of $100\,\Omega$. Removing the load yields an input impedance of $-j80\,\Omega$. With the unknown impedance attached, the input impedance is $(30 + j40)\Omega$. Determine the unknown impedance. $[\hat{Z}_L = 32 - j49\,\Omega]$

3.26 Shorted (or open-circuited) lengths of transmission lines can be constructed so that they appear at their input to be either capacitors or
inductors. This is how microwave circuit components are constructed.
Using discrete capacitors and inductors at these microwave frequencies
(GHz) would be useless due to the effect of their connection leads. To
demonstrate this, use the Smith chart to determine the length of a
transmission line (in meters) having a short-circuit load such that it
appears at its input terminals as a 30-nH inductor at 1 GHz. Assume a
microstripline ($v = 1.7 \times 10^8$ m/s) having a characteristic impedance
of 50 Ω. Verify your result using PSPICE. [3.55 cm]

3.27 An air-filled line ($v = 3 \times 10^8$ m/s) having a characteristic impedance
of 50 Ω is driven at a frequency of 30 MHz and is 1 m in length. The
line is terminated in a load of $\hat{Z}_L = (200 - j200)\Omega$. Determine
the input impedance using (**a**) the transmission-line model, and (**b**)
the approximate, lumped-pi model. [(a) $(12.89 - j51.49)\Omega$, (b)
$(13.76 - j52.25)\Omega$]

3.28 A coaxial cable ($v = 2 \times 10^8$ m/s) having a characteristic impedance
of 100 Ω is driven at a frequency of 4 MHz and is 5 m in length. The line
is terminated in a load of $\hat{Z}_L = (150 - j50)\Omega$ and the source is
$\hat{V}_S = 10\angle 0°$ V with $\hat{Z}_S = 25\ \Omega$. Determine the input and output
voltages to the line using (**a**) the transmission-line model, and (**b**) the
approximate lumped-pi model. [Exact: $\hat{V}(0) = 7.954\angle - 6.578°$,
$\hat{V}(\mathcal{L}) = 10.25\angle - 33.6°$;   approximate: $\hat{V}(0) = 7.959\angle - 5.906°$,
$\hat{V}(\mathcal{L}) = 10.27\angle - 35.02°$]

3.29 A low-loss coaxial cable has the following parameters:
$\hat{Z}_C \cong (75 + j0)\Omega$, $\alpha = 0.05$, $v = 2 \times 10^8$ m/s. Determine the input
impedance to a 11.175-m length of the cable at 400 MHz if the line is
terminated in (**a**) a short circuit, (**b**) an open circuit, and (**c**) a 300 Ω
resistor.   [(a) $90.209\angle - 34.86°\ \Omega$,   (b) $62.355\angle 34.06°\ \Omega$,   (c)
$66.7\angle 21.2°\ \Omega$]

3.30 Lossy cables, even low-loss cables, dissipate some power as signals
traverse them. Manufacturers of cables specify this loss at various
frequencies in dB/100 f or dB per some other linear dimension. The loss
is the ratio of the input power to the cable over the output power
extracted from it. In doing so, the cable is assumed to be matched so
there is only a forward-traveling wave (attenuated) on the line.
Determine an expression for the loss in such a cable in dB as a function
of the attenuation constant α. [Cable loss $= 8.686\alpha\ \mathcal{L}$ dB]

PART II

THREE-CONDUCTOR LINES
AND CROSSTALK

4

THE TRANSMISSION-LINE EQUATIONS FOR THREE-CONDUCTOR LINES

In Part II we investigate the prediction of crosstalk on three-conductor transmission lines. Crosstalk is the inadvertent coupling of a signal on one transmission line onto an adjacent transmission line, thereby having the possibility of creating interference in the modules that the adjacent line interconnect. In high-speed digital systems today, this has become a critical design problem preventing proper operation of those digital and analog systems. This is due to the electric fields **E** and magnetic fields **H** caused by the first line (the generator line) interacting with the second line (the receptor line), thereby inducing a voltage and a current in that line. The induced voltages at the ends of the receptor line are called the near-end and far-end crosstalk voltages, V_{NE} and V_{FE}, and can cause interference in the electronic devices that terminate the receptor line. In today's high-speed digital systems, crosstalk has become a very important factor in preventing them from working properly. Crosstalk is usually included under the general criterion of signal integrity.

4.1 THE TRANSMISSION-LINE EQUATIONS FOR THREE-CONDUCTOR LINES

Figure 4.1 shows the general configuration of a three-conductor transmission line that we will investigate.

Transmission Lines in Digital and Analog Electronic Systems: Signal Integrity and Crosstalk, By Clayton R. Paul
Copyright © 2010 John Wiley & Sons, Inc.

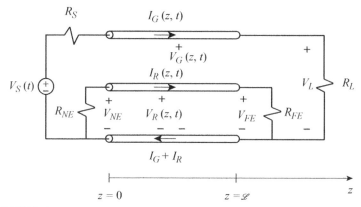

FIGURE 4.1. General three-conductor transmission line supporting crosstalk.

The *generator circuit* consists of the *loop* containing the source voltage, $V_S(t)$, the source resistance, R_S, the generator conductor (denoted as G), the load resistance, R_L, and the reference conductor (denoted as 0). The *receptor circuit* consists of the *loop* containing the near-end resistance, R_{NE}, the receptor conductor (denoted as R), the far-end resistance, R_{FE}, and the reference conductor. The third conductor, referred to as the *reference conductor*, is denoted as the zeroth (0) conductor and serves as (1) the reference for the voltages of the generator circuit, $V_G(z, t)$, and the receptor circuit, $V_R(z, t)$, and (2) the *return* for the generator circuit current, $I_G(z, t)$, and the receptor circuit current, $I_R(z, t)$. Note that the current in the reference conductor is $I_0 = -(I_G + I_R)$, so the net current along the system in the z direction is zero. All three conductors are assumed to be (1) parallel, (2) of uniform cross section along the line, and (3) of total length $\mathscr{L}$.

The current in the generator circuit, $I_G(z, t)$, creates *a magnetic field*, $B_G(z, t)$, about it. This magnetic field threads the generator circuit, giving a per-unit-length self-inductance of the generator circuit, l_G(H/m). A portion of this magnetic field also threads the receptor circuit, thereby inducing, by Faraday's law, a current $I_R(z, t)$ around the receptor circuit. Hence a per-unit-length mutual inductance exists between the generator and receptor circuits, l_m(H/m). Similarly, the current, $I_R(z, t)$, in the receptor circuit creates a *magnetic field*, $B_R(z, t)$, about it that threads the receptor circuit, giving a per-unit-length self-inductance of the receptor circuit, l_R(H/m). A portion of this magnetic field also threads the generator circuit, which is again represented as a per-unit-length mutual inductance between the receptor and generator circuits, l_m(H/m). (The mutual inductances between the two circuits are identical: $l_{GR} = l_{RG} \equiv l_m$.)

Similarly, the voltage of the generator circuit, $V_G(z, t)$, causes a per-unit-length charge, q_G, to be deposited on the generator circuit conductors which generates an *electric field* between the conductors of the generator circuit and between the conductors of the generator and receptor circuits. This *induces* a per-unit-length charge, q_R, on the conductors of the receptor circuit, resulting in a voltage, $V_R(z, t)$, being induced between the conductors of the receptor circuit. These charges also create electric fields between the generator and receptor circuits. Hence we have per-unit-length self-capacitances of each circuit, $c_G(\text{F/m})$ and $c_R(\text{F/m})$, along with a per-unit-length mutual capacitance, $c_m(\text{F/m})$, between the two circuits. (The mutual capacitances between the two circuits are equal: $c_{GR} = c_{RG} \equiv c_m$.)

Figure 4.2 shows the cross sections of typical three-conductor transmission lines composed of *wires*: three wires, two wires above a ground plane, and two wires within a circular cylindrical shield. The wires have radii r_w and circular cylindrical insulations of thicknesses t about them, having relative permittivities ε_r. Since these are dielectrics, they are nonmagnetic, having $\mu_r = 1$. In the first case, another wire serves as the reference conductor. In the second

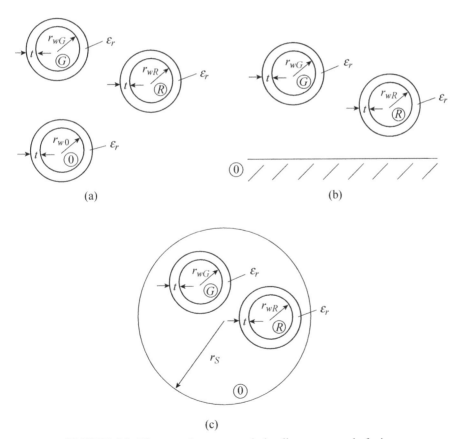

FIGURE 4.2. Three-conductor transmission lines composed of wires.

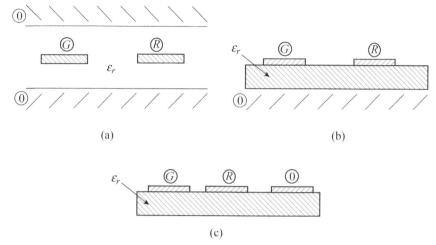

FIGURE 4.3. Three-conductor transmission lines composed of PCB lands.

case, an infinite ground plane serves as the reference conductor. In the third case, an overall shield serves as the reference conductor.

Figure 4.3 shows three-conductor lines that are found on (and in) PCBs. The first is a coupled stripline, the second is a coupled microstrip, and the third is an ordinary PCB not having inner planes. Note that the stripline is in a *homogeneous* medium, whereas the others are in an *inhomogeneous* medium, since the electric field is partly in free space and partly in the dielectric substrate.

Again the electric and magnetic fields lie in the transverse plane perpendicular to the line z axis. Hence the mode of propagation along the line is the Transverse ElectroMagnetic (TEM) mode of propagation. Hence the per-unit-length equivalent circuit is as shown in Fig. 4.4. Writing the KVL around each loop gives

$$
\begin{aligned}
V_G(z + \Delta z, t) - V_G(z, t) &= -l_G \, \Delta z \frac{\partial I_G(z, t)}{\partial t} \\
&\quad - l_m \, \Delta z \frac{\partial I_R(z, t)}{\partial t} \\
V_R(z + \Delta z, t) - V_R(z, t) &= -l_m \, \Delta z \frac{\partial I_G(z, t)}{\partial t} \\
&\quad - l_R \, \Delta z \frac{\partial I_R(z, t)}{\partial t}
\end{aligned}
\tag{4.1a}
$$

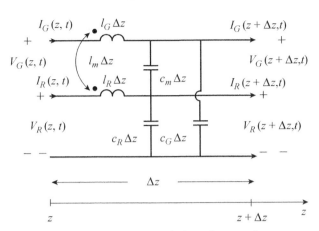

FIGURE 4.4. Per-unit-length equivalent circuit for a three-conductor transmission line.

Writing the KCL at each upper node gives

$$
I_G(z + \Delta z, t) - I_G(z, t) = -c_G \Delta z \frac{\partial V_G(z + \Delta z, t)}{\partial t}
$$
$$
- c_m \Delta z \frac{\partial (V_G(z + \Delta z, t) - V_R(z + \Delta z, t))}{\partial t}
$$
$$
I_R(z + \Delta z, t) - I_R(z, t) = -c_m \Delta z \frac{\partial (V_R(z + \Delta z, t) - V_G(z + \Delta z, t))}{\partial t}
$$
$$
- c_R \Delta z \frac{\partial V_R(z + \Delta z, t)}{\partial t}
$$

$$(4.1b)$$

Dividing both sides by Δz and letting $\Delta z \to 0$ gives the *multiconductor transmission-line (MTL) equations* as

$$
\frac{\partial V_G(z, t)}{\partial z} = -l_G \frac{\partial I_G(z, t)}{\partial t} - l_m \frac{\partial I_R(z, t)}{\partial t}
$$
$$
\frac{\partial V_R(z, t)}{\partial z} = -l_m \frac{\partial I_G(z, t)}{\partial t} - l_R \frac{\partial I_R(z, t)}{\partial t}
$$

$$(4.2a)$$

and

$$
\frac{\partial I_G(z, t)}{\partial z} = -(c_G + c_m) \frac{\partial V_G(z, t)}{\partial t} + c_m \frac{\partial V_R(z, t)}{\partial t}
$$
$$
\frac{\partial I_R(z, t)}{\partial z} = c_m \frac{\partial V_G(z, t)}{\partial t} - (c_R + c_m) \frac{\partial V_R(z, t)}{\partial t}
$$

$$(4.2b)$$

These are *coupled partial differential equations*. They can be written compactly in matrix form as

$$\frac{\partial}{\partial z}\mathbf{V}(z, t) = -\mathbf{L}\frac{\partial}{\partial t}\mathbf{I}(z, t)$$

$$\frac{\partial}{\partial z}\mathbf{I}(z, t) = -\mathbf{C}\frac{\partial}{\partial t}\mathbf{V}(z, t)$$

(4.3)

where the per-unit-length inductance and capacitance matrices are

$$\mathbf{L} = \begin{bmatrix} l_G & l_m \\ l_m & l_R \end{bmatrix} \quad \text{H/m}$$

(4.4a)

$$\mathbf{C} = \begin{bmatrix} c_G + c_m & -c_m \\ -c_m & c_R + c_m \end{bmatrix} \quad \text{(F/m)}$$

(4.4b)

and

$$\mathbf{V}(z, t) = \begin{bmatrix} V_G(z, t) \\ V_R(z, t) \end{bmatrix}$$

$$\mathbf{I}(z, t) = \begin{bmatrix} I_G(z, t) \\ I_R(z, t) \end{bmatrix}$$

(4.5)

The uncoupled second-order MTL equations are

$$\frac{\partial^2}{\partial z^2}\mathbf{V}(z, t) = \mathbf{LC}\frac{\partial^2}{\partial t^2}\mathbf{V}(z, t)$$

$$\frac{\partial^2}{\partial z^2}\mathbf{I}(z, t) = \mathbf{CL}\frac{\partial^2}{\partial t^2}\mathbf{I}(z, t)$$

(4.6)

The per-unit-length parameter matrices for lines in a *homogeneous medium* described by μ and ε are, like the case of two-conductor lines, related as

$$\mathbf{LC} = \mathbf{CL} = \underbrace{\mu\varepsilon}_{1/v^2} \mathbf{1}_2 \quad \text{homogeneous } medium$$

(4.7)

where $\mathbf{1}_2$ is the 2×2 identity matrix

$$\mathbf{1}_2 = \begin{bmatrix} 1 & 0 \\ 0 & 1 \end{bmatrix}$$

and v is the velocity of propagation of the TEM waves along the line:

$$v = \frac{1}{\sqrt{\mu\varepsilon}}$$

Hence for lines in a *homogeneous medium*, each per-unit-length parameter matrix can be found from the other:

$$\mathbf{L} = \frac{1}{v^2}\mathbf{C}^{-1} \qquad \text{homogeneous medium} \qquad (4.8a)$$

and

$$\mathbf{C} = \frac{1}{v^2}\mathbf{L}^{-1} \qquad \text{homogeneous medium} \qquad (4.8b)$$

For this case, $\mathbf{L}$ and $\mathbf{C}$ are 2×2 matrices, and the inverse of a 2×2 matrix can easily be obtained by hand as

$$\mathbf{M}^{-1} = \begin{bmatrix} a & b \\ c & d \end{bmatrix}^{-1}$$

$$= \frac{1}{ad - bc}\begin{bmatrix} d & -b \\ -c & a \end{bmatrix}$$

In other words, the inverse of a 2×2 matrix can be found by (1) swapping the main diagonal terms, (2) negating the off-diagonal terms, and (3) dividing each term by the determinant of the matrix, $ad - bc$.

For sinusoidal steady-state (phasor) excitation of the lines, we obtain by replacing $\partial/\partial t \Rightarrow j\omega$,

$$\frac{d}{dz}\hat{\mathbf{V}}(z) = -j\omega\mathbf{L}\hat{\mathbf{I}}(z)$$

$$\frac{d}{dz}\hat{\mathbf{I}}(z) = -j\omega\mathbf{C}\hat{\mathbf{V}}(z)$$

$$(4.9)$$

and the MTL equations become ordinary differential equations independent of time, t. The second-order uncoupled phasor differential equations become

$$\frac{d^2}{d^2z}\hat{\mathbf{V}}(z) = -\omega^2\mathbf{L}\mathbf{C}\hat{\mathbf{V}}(z)$$

$$\frac{d^2}{dz^2}\hat{\mathbf{I}}(z) = -\omega^2\mathbf{C}\mathbf{L}\hat{\mathbf{I}}(z)$$

$$(4.10)$$

We will spend our time solving these MTL equations. Their solution will be rather straightforward, although a bit more tedious than the solution for two-conductor lines.

4.2   THE PER-UNIT-LENGTH PARAMETERS

The entries in the per-unit-length parameter matrices, $\mathbf{L}$ and $\mathbf{C}$, are determined in the same fashion as for two-conductor lines, although the details are a bit more tedious. The per-unit-length inductance matrix, $\mathbf{L}$, relates the per-unit-length magnetic fluxes that thread the generator and receptor circuit *loops* to the currents on the conductors as

$$\begin{aligned}
\psi_G &= l_G I_G + l_m I_R \\
\psi_R &= l_m I_G + l_R I_R
\end{aligned} \tag{4.11a}$$

or

$$\boldsymbol{\psi} = \mathbf{LI} \tag{4.11b}$$

The per-unit-length capacitance matrix, $\mathbf{C}$, relates the per-unit-length electric charges on the generator and receptor circuit conductors to the voltages of each circuit as

$$\begin{aligned}
q_G &= (c_G + c_m)V_G - c_m V_R \\
q_R &= -c_m V_G + (c_R + c_m)V_R
\end{aligned} \tag{4.12a}$$

or

$$\mathbf{q} = \mathbf{CV} \tag{4.12b}$$

If we differentiate these with respect to time t, we get the MTL equations.
We can determine the individual per-unit-length parameters as

$$\begin{aligned}
l_G &= \left.\frac{\psi_G}{I_G}\right|_{I_R=0} \\[2mm]
l_R &= \left.\frac{\psi_R}{I_R}\right|_{I_G=0} \\[2mm]
l_m &= \left.\frac{\psi_G}{I_R}\right|_{I_G=0} \\[2mm]
&= \left.\frac{\psi_R}{I_G}\right|_{I_R=0}
\end{aligned} \tag{4.13a}$$

and

$$c_G + c_m = \left. \frac{q_G}{V_G} \right|_{V_R = 0}$$

$$c_R + c_m = \left. \frac{q_R}{V_R} \right|_{V_G = 0}$$

(4.13b)

$$c_m = - \left. \frac{q_G}{V_R} \right|_{V_G = 0}$$

$$= - \left. \frac{q_R}{V_G} \right|_{V_R = 0}$$

4.2.1 Wide-Separation Approximations for Wires

For these computations for wire-type lines we assume that (1) the wires are *widely separated*, and (2) are immersed in a *homogeneous medium* (such as air) having parameters $\mu = \mu_0$ and $\varepsilon = \varepsilon_r \varepsilon_0$. The restriction of wires being widely separated is not so restrictive as it may seem. We saw earlier that for wires that are very close together, the currents and charges on them will migrate toward the facing sides. Called the *proximity effect* this complicates the determination of the per-unit-length parameters. We also saw that the wide-separation assumption is reasonably valid for close separations, where the ratio of wire separation to wire radius is as small as $4 : 1$. In other words, one wire would just fit between two adjacent wires. Hence for practical applications this is not a very restrictive assumption.

For "widely separated" wires we may compute these per-unit-length parameters by assuming that the currents and the per-unit-length charges along the wires are distributed uniformly about the axes of the wires (i.e., the proximity effect is not well developed). Hence we can use the important results for the two fundamental subproblems developed earlier. The per-unit-length magnetic flux threading a surface that is parallel to the wire shown in Fig. 4.5 is

$$\psi = \frac{\mu_0 I}{2\pi} \ln \frac{R_2}{R_1} \quad \text{Wb/m} \qquad R_2 > R_1 \tag{4.14}$$

and the voltage between two points at distances R_1 and R_2 from the wire is obtained from Fig. 4.6 as

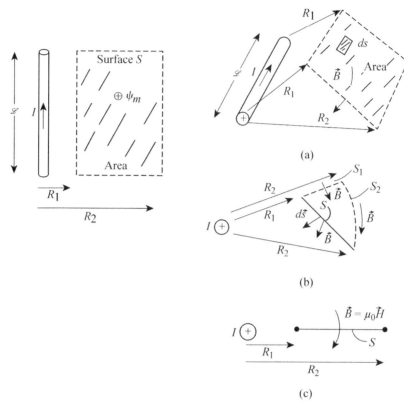

FIGURE 4.5. Fundamental subproblem for determining the per-unit-length inductances of wires.

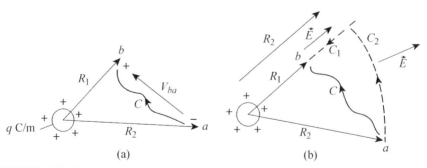

FIGURE 4.6. Fundamental subproblem for determining the per-unit-length capacitances of wires.

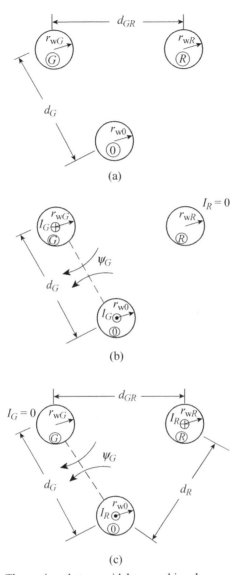

FIGURE 4.7. Three wires that are widely spaced in a homogeneous medium.

$$V = \frac{q}{2\pi\varepsilon} \ln \frac{R_2}{R_1} \quad \text{V} \qquad R_2 > R_1 \tag{4.15}$$

Three Wires   The first case is where the reference conductor is another wire. With reference to Fig. 4.7 and superimposing the fundamental subproblems above while assuming that the wires are widely spaced, so that the currents and per-unit-length charge distributions are uniformly distributed around the wire peripheries, the entries in the per-unit-length inductance matrix are

$$
\begin{aligned}
l_G &= \left.\frac{\psi_G}{I_G}\right|_{I_R=0} \\
&= \frac{\mu_0}{2\pi}\ln\frac{d_G}{r_{w0}} + \frac{\mu_0}{2\pi}\ln\frac{d_G}{r_{wG}} \\
&= \frac{\mu_0}{2\pi}\ln\frac{d_G^2}{r_{w0}r_{wG}}
\end{aligned}
\tag{4.16a}
$$

$$
\begin{aligned}
l_R &= \left.\frac{\psi_R}{I_R}\right|_{I_G=0} \\
&= \frac{\mu_0}{2\pi}\ln\frac{d_R}{r_{w0}} + \frac{\mu_0}{2\pi}\ln\frac{d_R}{r_{wR}} \\
&= \frac{\mu_0}{2\pi}\ln\frac{d_R^2}{r_{w0}r_{wR}}
\end{aligned}
\tag{4.16b}
$$

$$
\begin{aligned}
l_m &= \left.\frac{\psi_G}{I_R}\right|_{I_G=0} = \left.\frac{\psi_R}{I_G}\right|_{I_R=0} \\
&= \frac{\mu_0}{2\pi}\ln\frac{d_R}{d_{GR}} + \frac{\mu_0}{2\pi}\ln\frac{d_G}{r_{w0}} \\
&= \frac{\mu_0}{2\pi}\ln\frac{d_G d_R}{d_{GR}r_{w0}}
\end{aligned}
\tag{4.16c}
$$

Since the wires are assumed to be in a homogeneous medium, we can obtain the entries in the per-unit-length capacitance matrix from these as

$$
\mathbf{C} = \frac{\varepsilon_r}{v_0^2}\mathbf{L}^{-1}
$$

where the propagation velocity in this *homogeneous medium* characterized by μ_0 and $\varepsilon = \varepsilon_r\varepsilon_0$ is $v = 1/\sqrt{\mu_0\varepsilon_r\varepsilon_0}$.

Two Wires Above a Ground Plane   The next case is for two wires above an infinite, perfectly conducting ground plane. The ground plane is the reference

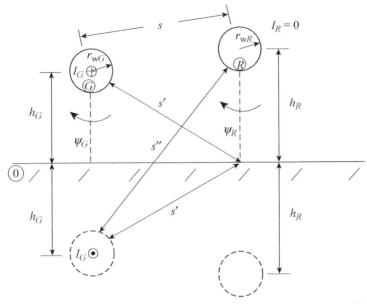

FIGURE 4.8. Two wires widely spaced in a homogeneous medium above an infinite ground plane.

conductor. With reference to Fig. 4.8 and superimposing the fundamental subproblems above while assuming that the wires are widely spaced, so that the currents and per-unit-length charge distributions are distributed uniformly around the wire peripheries, the entries in the per-unit-length inductance matrix are

$$
\begin{aligned}
l_G &= \left.\frac{\psi_G}{I_G}\right|_{I_R=0} \\
&= \frac{\mu_0}{2\pi}\ln\frac{h_G}{r_{wG}} + \frac{\mu_0}{2\pi}\ln\frac{2h_G}{h_G} \\
&= \frac{\mu_0}{2\pi}\ln\frac{2h_G}{r_{wG}}
\end{aligned}
\tag{4.17a}
$$

$$
\begin{aligned}
l_R &= \left.\frac{\psi_R}{I_R}\right|_{I_G=0} \\
&= \frac{\mu_0}{2\pi}\ln\frac{h_R}{r_{wR}} + \frac{\mu_0}{2\pi}\ln\frac{2h_R}{h_R} \\
&= \frac{\mu_0}{2\pi}\ln\frac{2h_R}{r_{wR}}
\end{aligned}
\tag{4.17b}
$$

and

$$
\begin{aligned}
l_m &= \left. \frac{\psi_R}{I_G} \right|_{I_R=0} = \left. \frac{\psi_G}{I_R} \right|_{I_G=0} \\[2mm]
&= \frac{\mu_0}{2\pi} \ln \frac{s'}{s} + \frac{\mu_0}{2\pi} \ln \frac{s''}{s'} \\[2mm]
&= \frac{\mu_0}{2\pi} \ln \frac{\sqrt{s^2 + 4h_G h_R}}{s^2} \\[2mm]
&= \frac{\mu_0}{4\pi} \ln \left(1 + \frac{4h_G h_R}{s^2} \right)
\end{aligned}
\tag{4.17c}
$$

Since the wires are assumed to be in a homogeneous medium, we can obtain the entries in the per-unit-length capacitance matrix from these as

$$
\mathbf{C} = \frac{\varepsilon_r}{v_0^2} \mathbf{L}^{-1}
$$

where the propagation velocity in this *homogeneous medium* characterized by μ_0 and $\varepsilon = \varepsilon_r \varepsilon_0$ is $v = 1/\sqrt{\mu_0 \varepsilon_r \varepsilon_0}$.

Two Wires Within an Overall Shield   The final case is for two wires located within an overall circular cylindrical shield. The shield is the reference conductor. The shield can be replaced with images located at distances r_s^2/d_G and r_s^2/d_R. With reference to Fig. 4.9 and superimposing the fundamental subproblems above while assuming that the wires are widely spaced, so that the currents and per-unit-length charge distributions are uniformly distributed around the wire peripheries, the entries in the per-unit-length inductance matrix are

$$
\begin{aligned}
l_G &= \left. \frac{\psi_G}{I_G} \right|_{I_R=0} \\[2mm]
&= \frac{\mu_0}{2\pi} \ln \frac{r_s - d_G}{r_{wG}} + \frac{\mu_0}{2\pi} \ln \frac{(r_s^2/d_G) - d_G}{(r_s^2/d_G) - r_s} \\[2mm]
&= \frac{\mu_0}{2\pi} \ln \frac{r_s^2 - d_G^2}{r_s r_{wG}}
\end{aligned}
\tag{4.18a}
$$

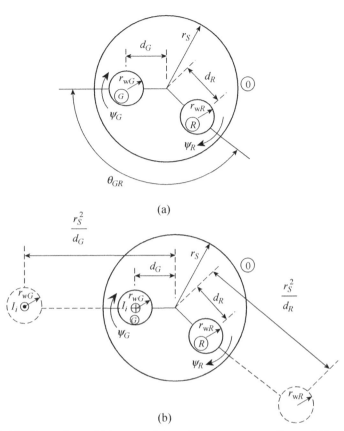

FIGURE 4.9. Two wires widely spaced in a homogeneous medium within an overall cylindrical shield.

$$l_R = \frac{\psi_R}{I_R}\bigg|_{I_G=0}$$

$$= \frac{\mu_0}{2\pi}\ln\frac{r_s - d_R}{r_{wR}} + \frac{\mu_0}{2\pi}\ln\frac{(r_s^2/d_R) - d_R}{(r_s^2/d_R) - r_s}$$

$$= \frac{\mu_0}{2\pi}\ln\frac{r_s^2 - d_R^2}{r_s r_{wR}}$$

(4.18b)

$$l_m = \frac{\psi_R}{I_G}\bigg|_{I_R=0} = \frac{\psi_G}{I_R}\bigg|_{I_G=0}$$

$$= \frac{\mu_0}{2\pi}\ln\left(\frac{d_R}{r_s}\sqrt{\frac{(d_G d_R)^2 + r_s^4 - 2d_G d_R r_s^2\cos\theta_{GR}}{(d_G d_R)^2 + d_R^4 - 2d_G d_R^3\cos\theta_{GR}}}\right)$$

(4.18c)

Since the wires are assumed to be in a homogeneous medium, we can obtain the entries in the per-unit-length capacitance matrix from these as

$$\boxed{\mathbf{C} = \frac{\varepsilon_r}{v_0^2}\mathbf{L}^{-1}}$$

where the propagation velocity in this *homogeneous medium* characterized by μ_0 and $\varepsilon = \varepsilon_r\varepsilon_0$ is $v = 1/\sqrt{\mu_0\varepsilon_r\varepsilon_0}$.

The WIDESEP Computer Program (WIDESEP.FOR, WIDESEP. EXE) WIDESEP.FOR is a FORTRAN program developed to compute the per-unit-length $\mathbf{L}$ and $\mathbf{C}$ matrices for widely separated wires. The program WIDESEP.EXE is the compiled and executable file. WIDESEP. EXE reads the input data stored in the ASCII file WIDESEP.IN and produces as output the file PUL.DAT, which contains the upper diagonals of the per-unit-length matrices $\mathbf{L}$ and $\mathbf{C}$. Although our concentration will be on three-conductor lines, the program handles n conductors, where n is any number up to the array dimensions of WIDESEP.FOR.

The locations of the wires for (1) three wires and (2) two wires above a ground plane are specified in a rectangular x–y coordinate system as shown in Fig. 4.10(a) and (b). For three wires, the reference wire is located at the origin of the coordinate system. For the case of two wires above a ground plane, the ground plane is the x axis of the coordinate system. In the case of two wires within an overall circular, cylindrical shield, the locations of the wires are specified in terms of the distance of the wire from the center of the shield and its angular location with reference to Fig. 4.10(c).

Following are three examples of the use of the WIDESEP.FOR (WIDESEP. EXE) program to compute the entries in the per-unit-length inductance and capacitance matrices $\mathbf{L}$ and $\mathbf{C}$ that are the upper diagonal output to the file PUL.DAT. Note that any such data overwrite any previous data written to the file PUL.DAT. There is only one input file to WIDESEP.FOR; it is is prepared using a standard ASCII editor and is called WIDESEP.IN.

As an example, the WIDESEP.IN data for three wires are

```
2       =NUMBER OF WIRES (EXCLUDING REFERENCE
          CONDUCTOR)
1       =REF CONDUCTOR (1=WIRE,2=GND PLANE,3=OVERALL
          SHIELD)
1.0     =RELATIVE PERMITTIVITY ER OF HOMOGENEOUS
            MEDIUM
```

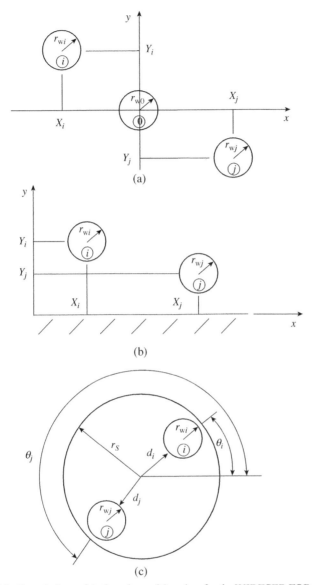

FIGURE 4.10. Descriptions of the locations of the wires for the WIDESEP. FOR program used in the input file WIDESEP.IN.

```
1.0          =RELATIVE PERMEABILITY MUR OF HOMOGENEOUS
             MEDIUM
7.5          =REF WIRE RADIUS OR SHIELD INTERIOR RADIUS
             (mils)
1            =WIRE #1
7.5          =RADIUS OF WIRE #1 (mils)
```

```
−1.27E-3 =    1=X1(meters),  2=H1(meters),  3=D1
              (mils)
0.       =    1=Y1(meters),  2=Y1(meters),  3=THETA1
              (degrees)
2             =WIRE #2
7.5           =RADIUS OF WIRE #2 (mils)
1.27E-3  =    1=X2(meters),  2=H2(meters),  3=D2
              (mils)
0.       =    1=Y2(meters),  2=Y2(meters),  3=THETA2
              (degrees)
```

The output data are in PUL.DAT:

```
1       1       7.58848E-07      =L(  1,   1)
1       2       2.40795E-07      =L(  1,   2)
2       2       7.58848E-07      =L(  2,   2)
1       1       1.63040E-11      =C(  1,   1)
1       2      −5.17352E-12      =C(  1,   2)
2       2       1.63040E-11      =C(  2,   2)
```

```
NUMBER OF WIRES= 2
RELATIVE PERMITTIVITY OF HOMOGENEOUS MEDIUM=   1.0
RELATIVE PERMEABILITY OF HOMOGENEOUS MEDIUM=   1.0
REFERENCE CONDUCTOR IS A WIRE WITH RADIUS (mils)=
   7.500E+00
```

WIRE	WIRE RADIUS	X COORDINATE	Y COORDINATE
#	(mils)	(meters)	(meters)
1	7.500E+00	−1.270E-03	0.000E+00
2	7.500E+00	1.270E-03	0.000E+00

The WIDESEP.IN data for two wires above a ground plane are

```
2         =NUMBER OF WIRES (EXCLUDING REFERENCE
           CONDUCTOR)
2         =REF CONDUCTOR (1=WIRE, 2=GND PLANE, 3=OVERALL
           SHIELD)
1.0       =RELATIVE PERMITTIVITY ER OF HOMOGENEOUS
```

```
            MEDIUM
1.0         =RELATIVE PERMEABILITY MUR OF HOMOGENEOUS
            MEDIUM
16.         =REF WIRE RADIUS OR SHIELD INTERIOR RADIUS
            (mils)
1           =WIRE #1
16.         =RADIUS OF WIRE #1 (mils)
2.E-2   =   1=X1(meters),  2=H1(meters),  3=D1(mils)
0.      =   1=Y1(meters),  2=Y1(meters),  3=THETA1
            (degrees)
2           =WIRE #2
16.         =RADIUS OF WIRE #2 (mils)
2.E-2   =   1=X2(meters),  2=H2(meters),  3=D2(mils)
2.E-2   =   1=Y2(meters),  2=Y2(meters),  3=THETA2
            (degrees)
```

The output data are in PUL.DAT:

```
1       1       9.17859E-07         =L(  1,   1)
1       2       1.60944E-07         =L(  1,   2)
2       2       9.17859E-07         =L(  2,   2)
1       1       1.25068E-11         =C(  1,   1)
1       2       -2.19302E-12        =C(  1,   2)
2       2       1.25068E-11         =C(  2,   2)
```

```
NUMBER OF WIRES= 2
RELATIVE PERMITTIVITY OF HOMOGENEOUS MEDIUM=  1.0
RELATIVE PERMEABILITY OF HOMOGENEOUS MEDIUM=  1.0
REFERENCE CONDUCTOR IS A GROUND PLANE
```

WIRE #	WIRE RADIUS (mils)	Y COORDINATE (meters)	HEIGHT ABOVE GND (meters)
1	1.600E+01	0.000E+00	2.000E-02
2	1.600E+01	2.000E-02	2.000E-02

The WIDESEP.IN data for two wires within an overall shield are

```
2           =NUMBER OF WIRES (EXCLUDING REFERENCE
            CONDUCTOR)
3           =REF CONDUCTOR (1=WIRE,2=GND PLANE,3=OVERALL
            SHIELD)
```

```
1.0        =RELATIVE PERMITTIVITY ER OF HOMOGENEOUS MEDIUM
1.0        =RELATIVE PERMEABILITY MUR OF HOMOGENEOUS
             MEDIUM
30.        =REF WIRE RADIUS OR SHIELD INTERIOR RADIUS
             (mils)
1          =WIRE #1
7.5        =RADIUS OF WIRE #1 (mils)
15.        =   1=X1(meters),  2=H1(meters),  3=D1(mils)
0.         =   1=Y1(meters),  2=Y1(meters),  3=THETA1
             (degrees)
2          =WIRE #2
7.5        =RADIUS OF WIRE #2 (mils)
15.        =   1=X2(meters),  2=H2(meters),  3=D2(mils)
180.       =   1=Y2(meters),  2=Y2(meters),  3=THETA2
             (degrees)
```

The output data are in PUL.DAT:

```
1        1        2.19722E-07      =L(  1,   1)
1        2        4.46287E-08      =L(  1,   2)
2        2        2.19722E-07      =L(  2,   2)
1        1        5.28179E-11      =C(  1,   1)
1        2       -1.07281E-11      =C(  1,   2)
2        2        5.28179E-11      =C(  2,   2)
```

```
NUMBER OF WIRES= 2
RELATIVE PERMITTIVITY OF HOMOGENEOUS MEDIUM=   1.0
RELATIVE PERMEABILITY OF HOMOGENEOUS MEDIUM=   1.0
REFERENCE CONDUCTOR IS A SHIELD WITH RADIUS (mils)=
  3.000E+01
```

WIRE #	WIRE RADIUS (mils)	POSITION RADIUS (mils)	POSITION ANGLE (degrees)
1	7.500E+00	1.500E+01	0.000E+00
2	7.500E+00	1.500E+01	1.800E+02

4.2.2   Numerical Methods

Computation of the entries in the per-unit-length inductance and capacitance matrices **L** and **C** for other structures is not possible in closed form and must be

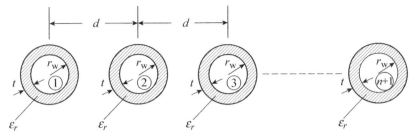

FIGURE 4.11. Common ribbon cable.

accomplished using numerical approximate methods that we describe in this section.

Wires with Dielectric Insulations (Ribbon Cables)   Ribbon cables are flat arrays of identical wires with insulation that has identical spacings (typically, 50 mils), as shown in Fig. 4.11. Typical dimensions and properties of common ribbon cables are

$$d = 50 \text{ mils}$$
$$r_w = 7.5 \text{ mils (No. 20 gauge, } 7 \times 36)$$
$$t = 10 \text{ mils}$$
$$\varepsilon_r = 3.5 \text{ (PVC)}$$

There are two important properties that prevent using the previous wide-separation formulas: (1) The metallic wires are surrounded by dielectric insulations, hence this is an inhomogeneous medium. (2) The ratio of separation to overall radius of the wires (wire radius + dielectric thickness) is 2.86, which means that the proximity effect will be pronounced and we cannot make the assumption that the per-unit-length charges will be distributed uniformly around the wire and dielectric peripheries. To handle these problems, we note that the dielectric insulations will have *bound-charge distributions* around their inner *and* outer surfaces, $\rho_b(\text{C/m}^2)$, while the surfaces of the metallic wires will have *free-charge distributions* around them, $\rho_f(\text{C/m}^2)$.

As an example, consider the parallel-plate capacitor illustrated in Fig. 4.12. When we attach the voltage source, *free charge* will flow from the battery to the capacitor plates. When we insert a block of dielectric between the plates, the microscopic *dipoles* in the dielectric will align with the **E** field, thereby inducing *bound charges* on the faces of the dielectric. This will cause more *free charge* to flow from the battery to the plates. Hence the capacitance (which is the ratio of *free charge* to the voltage between the two plates) will increase.

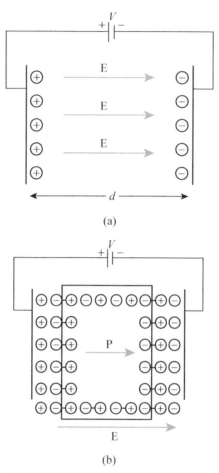

FIGURE 4.12. Parallel-plate capacitor and bound charge formed on the surfaces of the dielectric.

For the ribbon cable, we expand these distributions as Fourier series in terms of the angle around the surfaces. Around the outer surface of the dielectric insulation we represent the bound charge on its surface with a Fourier series as

$$\hat{\rho}_{ib} = \hat{\alpha}_{i0} + \sum_{k=1}^{\hat{N}_i} \hat{\alpha}_{ik}\cos k\theta_i \qquad \text{C/m}^2 \qquad (4.19)$$

where the subscript i is for the ith wire. Around the outer surface of the metallic wire (the inner surface of the dielectric insulation) we have both free and bound charge. The *total* bound charge on the inner and outer surfaces of the dielectric insulation will be equal but opposite in sign. Because of the different radii of the surfaces, their distributions will be different. Similarly, we represent the bound and free charge on the conductor–dielectric surface with a Fourier series as

$$\rho_{if} - \rho_{ib} = \alpha_{i0} + \sum_{k=1}^{N_i} \alpha_{ik} \cos k\theta_i \qquad (C/m^2) \qquad (4.20)$$

At the air–dielectric surface, the total per-unit-length bound charge is

$$q_{ib} = \int_{\theta_i=0}^{2\pi} \hat{\rho}_{ib}(r_w + t)d\theta_i$$

$$= 2\pi(r_w + t)\hat{\alpha}_{i0} \qquad C/m \qquad (4.21)$$

At the conductor–dielectric surface, the total per-unit-length free plus bound charge is

$$q_{if} - q_{ib} = \int_{\theta_i=0}^{2\pi} (\rho_{if} - \rho_{ib})r_w d\theta_i$$

$$= 2\pi r_w \alpha_{i0} \qquad C/m \qquad (4.22)$$

Hence the total per-unit-length free charge around the metallic wire is

$$\boxed{q_{if} = 2\pi(r_w + t)\hat{\alpha}_{i0} + 2\pi r_w \alpha_{i0} \qquad C/m} \qquad (4.23).$$

Each wire has associated with it a total of $N_i + 1 + \hat{N}_i + 1$ unknown expansion coefficients. To determine these, we select $N_i + 1$ match points around the wire periphery where we impose the condition that the total voltage there due to all the charge distribtutions will equal the voltage of that wire. We select $\hat{N}_i + 1$ match points around the air–dielectric periphery, where we impose the condition that the components of the **D** fields due to all the charge distribtutions on either side of the boundary that are normal to the boundary will be equal. This gives a set of simultaneous equations to be solved for the α_{i0} and $\hat{\alpha}_{i0}$ average-value coefficients for each wire, giving the per-unit-length capacitance matrix, **C**, for a chosen reference conductor. The per-unit-length inductance matrix can be found from the capacitance matrix with the dielectric insulations removed, $\mathbf{C}_0$, as $\mathbf{L} = 1/v_0^2\mathbf{C}_0^{-1}$.

The RIBBON Computer Program (RIBBON.FOR, RIBBON. EXE)
RIBBON.FOR is a FORTRAN program developed to compute the per-unit-length **L** and **C** matrices for ribbon cables. The program RIBBON. EXE is the compiled and executable file. RIBBON.EXE reads the input data

stored in the file RIBBON.IN and produces as output the file PUL.DAT, which contains the upper diagonals of the per-unit-length matrices **L** and **C**. (Note that any data in a previous PUL.DAT are overwritten by the new data.) Although our concentration will be on three-conductor lines, the program handles $n + 1$ wires, where $n + 1$ is any number up to the array dimensions of RIBBON.FOR. The number of the reference wire is chosen in RIBBON.IN. Note that the wires are numbered *left to right* as $1, 2, \ldots, n + 1$. When the reference wire is chosen (0), the wires are again numbered sequentially *from left to right*, omitting the reference wire chosen, as $1, 2, \ldots, n$.

For example, the RIBBON.IN data for three wires are

```
3         =TOTAL NUMBER OF WIRES
20        =NUMBER OF FOURIER COEFFICIENTS
2         =NUMBER OF REFERENCE WIRE
7.5       =WIRE RADIUS (mils)
10.0      =INSULATION THICKNESS (mils)
3.5       =RELATIVE DIELECTRIC CONSTANT OF INSULATION
50.0      =ADJACENT WIRE SEPARATION (mils)
```

The output data are in PUL.DAT:

```
1       1          7.48501E-07      =L (  1,   1)
1       2          2.40801E-07      =L (  1,   2)
2       2          7.48501E-07      =L (  2,   2)
1       1          2.49819E-11      =C (  1,   1)
1       2         -6.26613E-12      =C (  1,   2)
2       2          2.49819E-11      =C (  2,   2)
1       1          1.65812E-11      =C0 ( 1,   1)
1       2         -5.33434E-12      =C0 ( 1,   2)
2       2          1.65812E-11      =C0 ( 2,   2)
```

```
NUMBER OF WIRES=   3
NUMBER OF FOURIER COEFFICIENTS=   20
REFERENCE WIRE=   2
WIRE RADIUS (mils)=   7.500E+00
DIELECTRIC INSULATION THICKNESS (mils)=   1.000E+01
DIELECTRIC CONSTANT OF INSULATION=   3.500E+00
CENTER-TO-CENTER SEPARATION (mils)= 5.000E+01
```

Conductors Having Rectangular Cross Sections (PCB Lands)   The conductors (lands) of PCBs have rectangular cross sections. The thickness of typical lands (etched from 1-oz copper) is 1.4 mils, and the typical land widths

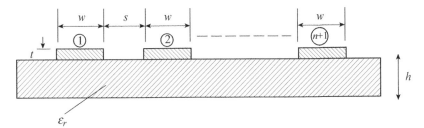

FIGURE 4.13. Land numbering for the PCB.FOR program.

range from 5 to 25 mils. To simplify the calculation we assume that the land thicknesses are zero ($t = 0$). Typical thicknesses of the boards are 47 and 64 mils. The supporting board is made of glass-epoxy (FR-4) and has a relative permittivity of about $\varepsilon_r = 4.7$. Again note that the lands are numbered *left to right* as $1, 2, \ldots, n + 1$, as illustrated in Fig. 4.13. When the reference land is chosen (0), the lands are again numbered *from left to right*, omitting the reference land chosen, as $1, 2, \ldots, n$.

Numerical methods must again be used for computing the per-unit-length inductances and capacitances of PCB lands because the charge and current tends to *peak at the edges of the land cross sections*, as illustrated in Fig. 4.14.

To determine the per-unit-length charge on each land, we model the charge distribution with piecewise-constant segments whose heights, α_{ik} (C/m^2), are unknown. Then the per-unit-length charge on the ith land is

$$q_i = \sum_{k=1}^{N_i} \alpha_{ik} w_{ik} \qquad \text{C/m} \qquad (4.24)$$

To determine the unknown distribution coefficients, α_{ik}, we again choose N_i match points at the midpoints of each of the subsections of each land, at which we enforce the requirement that the voltage at the midpoint of that subsection that is due to all the charges on all the land subsections is equal to the voltage of the land. This again gives $(n + 1)N_i$ equations in the $(n + 1)N_i$ unknowns. We incorporate the inhomogeneous medium by incorporating the board into this using images. Solving these for the α_{ik}, we can determine the per-unit-length capacitance matrix with ($\mathbf{C}$) and without ($\mathbf{C}_0$) the board present. Then the per-unit-length inductance matrix can be found from $\mathbf{L} = (1/v_0^2)\mathbf{C}_0^{-1}$.

The PCB Computer Program (PCB.FOR, PCB.EXE)   PCB.FOR is a FORTRAN program developed to compute the per-unit-length $\mathbf{L}$ and $\mathbf{C}$ matrices for the PCB shown in Fig. 4.13. The program PCB.EXE is the compiled and executable file. PCB.EXE reads the input data stored in the file PCB.IN and produces as output the file PUL.DAT, which contains the upper diagonals of the per-unit-length matrices $\mathbf{L}$ and $\mathbf{C}$. (Note that any data in a

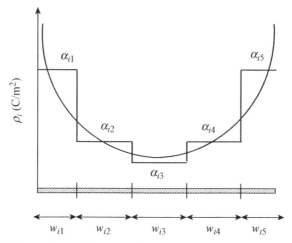

FIGURE 4.14. Peaking of the charge distributions across the lands at the edges.

previous PUL.DAT are overwritten by the new data.) Although our concentration will be on three-conductor lines, the program handles $n + 1$ conductors, where $n + 1$ is any number up to the array dimensions of PCB. FOR.

As an example, the PCB.IN data for a three-land PCB are

```
3         =TOTAL NUMBER OF LANDS
30        =NUMBER OF CONDUCTOR SUBSECTIONS
3         =NUMBER OF REFERENCE LAND
15.0      =LAND WIDTH (mils)
45.0      =EDGE-TO-EDGE LAND SEPARATION (mils)
47.       =BOARD THICKNESS (mils)
4.7       =BOARD RELATIVE DIELECTRIC CONSTANT
```

The output data are in PUL.DAT:

```
1     1        1.38315E-06      =L (  1,   1)
1     2        6.91573E-07      =L (  1,   2)
2     2        1.10707E-06      =L (  2,   2)
1     1        2.96949E-11      =C (  1,   1)
1     2       -2.02619E-11      =C (  1,   2)
2     2        4.05238E-11      =C (  2,   2)
1     1        1.16982E-11      =C0 (  1,   1)
1     2       -7.30774E-12      =C0 (  1,   2)
2     2        1.46155E-11      =C0 (  2,   2)
```

```
NUMBER OF LANDS=    3
NUMBER OF DIVISIONS PER LAND=    30
REFERENCE LAND=    3
LAND WIDTH (mils)=    1.500E+01
EDGE-TO-EDGE SEPARATION (mils)=    4.500E+01
BOARD THICKNESS (mils)=    4.700E+01
RELATIVE DIELECTRIC CONSTANT=    4.700E+00
```

The MSTRP Computer Program (MSTRP.FOR, MSTRP.EXE)   MSTRP.
FOR is a FORTRAN program developed to compute the per-unit-length L and
C matrices for the coupled microstripline shown in Fig. 4.3(b), where there is a
ground plane (the reference conductor) beneath the board. The program
MSTRP.EXE is the compiled and executable file. MSTRP.EXE reads the
input data stored in the file MSTRP.IN and produces as output the file PUL.
DAT, which contains the upper diagonals of the per-unit-length matrices L and
C. (Note that any data in a previous PUL.DAT are overwritten by the new data.)
Although our concentration will be on three-conductor lines (two lands and
the ground plane), the program handles *n* conductors, where *n* is any number
up to the array dimensions of MSTRP.FOR.

**As an example, the MSTRP.IN data for a two-land PCB above a ground
plane–coupled microstrip are**

```
2       =TOTAL NUMBER OF LANDS (EXCLUSIVE OF GND PLANE)
30      =NUMBER OF CONDUCTOR SUBSECTIONS
100.    =LAND WIDTH (mils)
100.    =EDGE-TO-EDGE LAND SEPARATION (mils)
62.     =BOARD THICKNESS (mils)
4.7     =BOARD RELATIVE DIELECTRIC CONSTANT
```

The output data are in PUL.DAT:

```
1       1       3.35327E-07     =L(  1,   1)
1       2       3.71527E-08     =L(  1,   2)
2       2       3.35327E-07     =L(  2,   2)
1       1       1.15511E-10     =C(  1,   1)
1       2      -4.92724E-12     =C(  1,   2)
2       2       1.15511E-10     =C(  2,   2)
1       1       3.35934E-11     =C0(  1,   1)
1       2      -3.72200E-12     =C0(  1,   2)
2       2       3.35934E-11     =C0(  2,   2)
```

```
NUMBER OF LANDS=    2
NUMBER OF DIVISIONS PER LAND=   30
LAND WIDTH (mils) =   1.000E+02
EDGE-TO-EDGE SEPARATION (mils) =   1.000E+02
BOARD THICKNESS (mils) =   6.200E+01
RELATIVE DIELECTRIC CONSTANT=   4.700E+00
```

The STRPLINE Computer Program (STRPLINE.FOR, STRPLINE. EXE)

STRPLINE.FOR is a FORTRAN program developed to compute the per-unit-length **L** and **C** matrices for the coupled stripline shown in Fig. 4.3 (a), where there is a ground plane (the reference conductor) on both sides of the board and the lands are sandwiched between them. Note that unlike the ribbon cable, the PCB, and the coupled microstrip line, this is a line in a *homogeneous medium*. The program STRPLINE.EXE is the compiled and executable file. STRPLINE.EXE reads the input data stored in the file STRPLINE.IN and produces as output the file PUL.DAT, which contains the upper diagonals of the per-unit-length matrices **L** and **C**. (Note that any data in a previous PUL. DAT are overwritten by the new data.) Although our concentration will be on three-conductor lines (two lands and the ground planes), the program handles *n* conductors, where *n* is any number up to the array dimensions of STRPLINE.FOR.

As an example, the STRPLINE.IN data for a two-land PCB between ground planes–coupled striplines are

```
2         =TOTAL NUMBER OF LANDS (EXCLUSIVE OF GND PLANES)
30        =NUMBER OF CONDUCTOR SUBSECTIONS
5.        =LAND WIDTH (mils)
5.        =EDGE-TO-EDGE LAND SEPARATION (mils)
20.       =SEPARATION BETWEEN GROUND PLANES (mils)
4.7       =BOARD RELATIVE DIELECTRIC CONSTANT
```

The output data are in PUL.DAT:

```
1       1        4.63055E-07      =L(  1,   1)
1       2        9.21843E-08      =L(  1,   2)
2       2        4.63054E-07      =L(  2,   2)
1       1        1.17594E-10      =C(  1,   1)
1       2       -2.34105E-11      =C(  1,   2)
2       2        1.17594E-10      =C(  2,   2)
1       1        2.50201E-11      =C0(  1,   1)
```

```
1        2       -4.98097E-12        =C0( 1,   2)
2        2        2.50201E-11        =C0( 2,   2)
```

NUMBER OF LANDS=   2
NUMBER OF DIVISIONS PER LAND=   30
LAND WIDTH (mils)=   5.000E+00
EDGE-TO-EDGE SEPARATION (mils)=   5.000E+00
SEPARATION BETWEEN GROUND PLANES (mils)=   2.000E+01
RELATIVE DIELECTRIC CONSTANT=   4.700E+00

PROBLEMS

4.1 In Chapter 2 we determined the exact and approximate values for the per-unit-length capacitance and inductance of a two-wire line (typical of ribbon cables used to interconnect electronic components) whose wires have radii of 7.5 mils (0.19 mm) and are separated by 50 mils (1.27 mm). [Exact: 14.8 pF/m and 0.75 μH/m; approximate: 14.6 pF/m and 0.759 μH/m] Repeat this calculation using WIDESEP.FOR.

4.2 In Chapter 2 we determined the exact and approximate values for the per-unit-length capacitance and inductance of one wire of radius 16 mils (0.406 mm) at a height of 100 mils (2.54 mm) above a ground plane. [Exact: 22.1 pF/m and 0.504 μH/m; approximate: 22.0 pF/m and 0.505 μH/m] Repeat this calculation using WIDESEP.FOR.

4.3 In Chapter 2 we determined the per-unit-length capacitance and inductance of a coaxial cable (RG-58U) where the inner wire has radius 16 mils (0.406 mm) and the shield has an inner radius of 58 mils (1.47 mm). The dielectric is polyethylene with a relative permittivity of 2.3. [99.2 pF/m and 0.258 μH/m] Repeat this calculation using WIDESEP.FOR. Can you figure out why the result of WIDESEP.FOR for the per-unit-length capacitance c is not exactly the same as the result from the formula derived previously? Shouldn't they be identical?

4.4 Consider a three-wire ribbon cable whose wires have radii of 7.5 mils and are separated by 50 mils. The wire insulations have thicknesses of 10 mils, and the relative permittivity is 3.5. One of the outer wires is the reference wire. Use RIBBON.FOR to compute the entries in **C** and **L**. Sanity-check these results. First use five Fourier coefficients and then increase these to 10 to check the convergence of the results. Compare these "exact" results to the wide-separation results obtained with WIDESEP.FOR. Are the entries in **L** approximately the same? Are the entries in **C** approximately the same? If not, why not?

4.5   Consider a three-wire ribbon cable whose wires have radii of 7.5 mils and are separated by 50 mils. The wire insulations have thicknesses of 10 mils, and the relative permittivity is 3.5. The inner wire is the reference wire. Use RIBBON.FOR to compute the entries in **C** and **L**. Sanity-check these results. First use five Fourier coefficients and then increase these to 10 to check the convergence of the results. Compare these exact results to the wide-separation results obtained with WIDE-SEP.FOR. Are the entries in **L** approximately the same? Are the entries in **C** approximately the same? If not, why not?

4.6   In Chapter 2 we determined the per-unit-length capacitance and inductance of a one-conductor stripline with dimensions $s = 20$ mils (0.508 mm), $w = 5$ mils (0.127 mm), and $\varepsilon_r = 4.7$. [113.2 pF/m and 0.461 µH/m] Determine these using STRPLINE.FOR.

4.7   In Chapter 2 we determined the per-unit-length capacitance and inductance of a one-conductor microstripline with dimensions $h = 50$ mils (1.27 mm), $w = 5$ mils (0.127 mm), and $\varepsilon_r = 4.7$. [38.46 pF/m and 0.877 µH/m. The effective relative permittivity is $\varepsilon'_r = 3.034$.] Determine these using MSTRP.FOR.

4.8   In Chapter 2 we determined the per-unit-length capacitance and inductance of a two-conductor PCB with dimensions $s = 15$ mils (0.381 mm), $w = 15$ mils (0.381 mm), $h = 62$ mils (1.575 mils), and $\varepsilon_r = 4.7$. [38.53 pF/m and 0.804 µH/m. The effective relative permittivity is $\varepsilon'_r = 2.787$.] Determine these using PCB.FOR.

5

SOLUTION OF THE TRANSMISSION-LINE EQUATIONS FOR THREE-CONDUCTOR LOSSLESS LINES

In this chapter we determine a simple method for solution of the transmission-line equations for a three-conductor *lossless line* and the resulting crosstalk, where we neglect losses in the conductors and in the surrounding dielectric. Solution of the transmission-line equations for any *lossless multiconductor line* is virtually *trivial*, as we will see in this chapter. A SPICE (PSPICE) subcircuit model will be developed for any *lossless* MTL having any number of parallel conductors. Terminations (linear or nonlinear) can then be attached to that PSPICE subcircuit model, and the resulting crosstalk can be determined easily and accurately.

For *three-conductor lossless transmission lines* the first-order transmission-line equations that describe crosstalk between the two transmission lines are *coupled* and of the form

$$\frac{\partial \mathbf{V}(z,t)}{\partial z} = -\mathbf{L}\frac{\partial \mathbf{I}(z,t)}{\partial t} \qquad (5.1a)$$

$$\frac{\partial \mathbf{I}(z,t)}{\partial z} = -\mathbf{C}\frac{\partial \mathbf{V}(z,t)}{\partial t} \qquad (5.1b)$$

Transmission Lines in Digital and Analog Electronic Systems: Signal Integrity and Crosstalk, By Clayton R. Paul
Copyright © 2010 John Wiley & Sons, Inc.

where the 2×1 vectors of line voltages and line currents have entries that are functions of position along the line, z, and time, t:

$$\mathbf{V}(z, t) = \begin{bmatrix} V_G(z, t) \\ V_R(z, t) \end{bmatrix} \tag{5.2a}$$

$$\mathbf{I}(z, t) = \begin{bmatrix} I_G(z, t) \\ I_R(z, t) \end{bmatrix} \tag{5.2b}$$

The entries in the 2×2 per-unit-length inductance and capacitance matrices, **L** and **C**, respectively, are *independent* of position along the line, z, and time, t and contain all the cross-sectional dimensions and properties of the line. These are of the form

$$\mathbf{L} = \begin{bmatrix} l_G & l_m \\ l_m & l_R \end{bmatrix} \tag{5.3a}$$

and

$$\mathbf{C} = \begin{bmatrix} c_G + c_m & -c_m \\ -c_m & c_R + c_m \end{bmatrix} \tag{5.3b}$$

5.1 DECOUPLING THE TRANSMISSION-LINE EQUATIONS WITH MODE TRANSFORMATIONS

The key to solving the *coupled* MTL equations in (5.1) is to *decouple them simultaneously* with *similarity transformations*. Define a *change of variables* to convert the *actual* line voltages and *actual* line currents to the *mode* voltages and *mode* currents as

$$\boxed{\mathbf{V}(z, t) = \mathbf{T}_V \mathbf{V}_{\mathrm{mode}}(z, t)} \tag{5.4a}$$

$$\boxed{\mathbf{I}(z, t) = \mathbf{T}_I \mathbf{I}_{\mathrm{mode}}(z, t)} \tag{5.4b}$$

Observe that the 2×2 mode transformations $\mathbf{T}_V$ and $\mathbf{T}_I$ are *independent* of position along the line, z, and time, t. These vectors of *mode voltages* and *mode currents* are of the form

$$\mathbf{V}_{\mathrm{mode}}(z, t) = \begin{bmatrix} V_{mG}(z, t) \\ V_{mR}(z, t) \end{bmatrix} \tag{5.5a}$$

$$\mathbf{I}_{\mathrm{mode}}(z, t) = \begin{bmatrix} I_{mG}(z, t) \\ I_{mR}(z, t) \end{bmatrix} \quad (5.5b)$$

and the 2×2 mode tranformations have entries

$$\mathbf{T}_V = \begin{bmatrix} T_{VGG} & T_{VGR} \\ T_{VRG} & T_{VRR} \end{bmatrix} \quad (5.6a)$$

$$\mathbf{T}_I = \begin{bmatrix} T_{IGG} & T_{IGR} \\ T_{IRG} & T_{IRR} \end{bmatrix} \quad (5.6b)$$

Now substitute the mode transformations in (5.4) into the MTL equations to give

$$\boxed{\frac{\partial \mathbf{V}_{\mathrm{mode}}}{\partial z} = - \underbrace{\mathbf{T}_V^{-1} \mathbf{L} \mathbf{T}_I}_{\mathbf{L}_m} \frac{\partial \mathbf{I}_{\mathrm{mode}}}{\partial t}} \quad (5.7a)$$

$$\boxed{\frac{\partial \mathbf{I}_{\mathrm{mode}}}{\partial z} = - \underbrace{\mathbf{T}_I^{-1} \mathbf{C} \mathbf{T}_V}_{\mathbf{C}_m} \frac{\partial \mathbf{V}_{\mathrm{mode}}}{\partial t}} \quad (5.7b)$$

If we can choose the 2×2 mode transformations $\mathbf{T}_V$ and $\mathbf{T}_I$ such that $\mathbf{L}$ and $\mathbf{C}$ are *simultaneously* diagonalized as

$$\begin{aligned} \mathbf{L}_m &= \mathbf{T}_V^{-1} \mathbf{L} \mathbf{T}_I \\ &= \begin{bmatrix} l_{mG} & 0 \\ 0 & l_{mR} \end{bmatrix} \end{aligned} \quad (5.8a)$$

and

$$\begin{aligned} \mathbf{C}_m &= \mathbf{T}_I^{-1} \mathbf{C} \mathbf{T}_V \\ &= \begin{bmatrix} c_{mG} & 0 \\ 0 & c_{mR} \end{bmatrix} \end{aligned} \quad (5.8b)$$

the MTL equations are decoupled into two sets of *uncoupled* two-conductor mode lines as

$$\begin{aligned} \frac{\partial V_{mG}(z, t)}{\partial z} &= - l_{mG} \frac{\partial I_{mG}(z, t)}{\partial t} \\ \frac{\partial I_{mG}(z, t)}{\partial z} &= - c_{mG} \frac{\partial V_{mG}(z, t)}{\partial t} \end{aligned} \quad (5.9a)$$

and

$$\frac{\partial V_{mR}(z, t)}{\partial z} = -l_{mR}\frac{\partial I_{mR}(z, t)}{\partial t}$$
$$\frac{\partial I_{mR}(z, t)}{\partial z} = -c_{mR}\frac{\partial V_{mR}(z, t)}{\partial t} \tag{5.9b}$$

giving two *uncoupled* and independent sets of two-conductor mode transmission-line equations having characteristic impedances of *each mode* as

$$Z_{CmG} = \sqrt{\frac{l_{mG}}{c_{mG}}}$$
$$Z_{CmR} = \sqrt{\frac{l_{mR}}{c_{mR}}} \tag{5.10}$$

and velocities of propagation of *each mode* as

$$v_{mG} = \frac{1}{\sqrt{l_{mG}c_{mG}}}$$
$$v_{mR} = \frac{1}{\sqrt{l_{mR}c_{mR}}} \tag{5.11}$$

Note that *each of these decoupled mode transmission lines* can be represented by a two-conductor line and can be simulated using the built-in PSPICE two-conductor line model! From now on, we omit showing the dependence of the line and mode voltages and currents on position along the line, z, and time, t, with that implicit understanding.

5.2 THE SPICE SUBCIRCUIT MODEL

But how do we get back from the mode voltages and currents to the actual line voltages and currents? The answer is to build a PSPICE model using *controlled sources* that simulates these mode transformations. The mode transformations in (5.4) can be written as

$$\mathbf{V} = \mathbf{T}_V \mathbf{V}_{\text{mode}} \tag{5.12a}$$

$$\mathbf{I}_{\text{mode}} = \mathbf{T}_I^{-1}\mathbf{I} \tag{5.12b}$$

or

$$V_G = T_{VGG}V_{mG} + T_{VGR}V_{mR}$$
$$V_R = T_{VRG}V_{mG} + T_{VRR}V_{mR} \tag{5.13a}$$

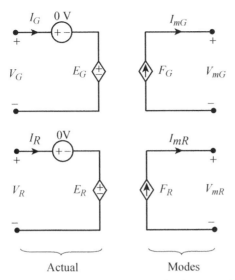

FIGURE 5.1. Equivalent circuit giving the transformation between the actual line voltages and currents and the mode voltages and currents.

and

$$I_{mG} = T_{IGG}^{-1}I_G + T_{IGR}^{-1}I_R$$
$$I_{mR} = T_{IRG}^{-1}I_G + T_{IRR}^{-1}I_R \tag{5.13b}$$

Note that $T_{IGG}^{-1} \neq 1/T_{IGG}$; it just symbolizes the entry in $\mathbf{T}_I^{-1}$. The equivalent circuit that gives this transformation is shown in Fig. 5.1, where

$$E_G = T_{VGG}V_{mG} + T_{VGR}V_{mR}$$
$$E_R = T_{VRG}V_{mG} + T_{VRR}V_{mR} \tag{5.14a}$$

and

$$F_G = T_{IGG}^{-1}I_G + T_{IGR}^{-1}I_R$$
$$F_R = T_{IRG}^{-1}I_G + T_{IRR}^{-1}I_R \tag{5.14b}$$

The complete PSPICE subcircuit model is shown in Fig. 5.2.

But how do we determine the mode transformations $\mathbf{T}_V$ and $\mathbf{T}_I$ that will *simultaneously diagonalize* $\mathbf{L}$ and $\mathbf{C}$? For 2×2 $\mathbf{L}$ and $\mathbf{C}$, this is simple. There are some important, basic facts about similarity transformations of matrices.

1. $\mathbf{L}$ and $\mathbf{C}$ are both real and symmetric; that is, their entries are real numbers and $L_{12} = L_{21}$ and $C_{12} = C_{21}$.

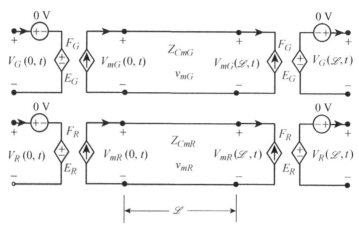

FIGURE 5.2. Complete PSPICE subcircuit model.

2. Any real symmetric matrix

$$\mathbf{M} = \begin{bmatrix} M_{11} & M_{12} \\ M_{12} & M_{22} \end{bmatrix}$$

can be diagonalized with a real *orthogonal transformation*

$$\mathbf{T} = \begin{bmatrix} T_{11} & T_{12} \\ T_{21} & T_{22} \end{bmatrix}$$

such that

$$\begin{aligned} \mathbf{T}^{-1}\mathbf{MT} &= \mathbf{T}^{t}\mathbf{MT} \\ &= \begin{bmatrix} D_1 & 0 \\ 0 & D_2 \end{bmatrix} \end{aligned} \tag{5.15}$$

where the inverse of **T** is its *transpose*:

$$\mathbf{T}^{-1} = \mathbf{T}^{t} \tag{5.16}$$

and $\mathbf{T}^{t}$ denotes its *transpose* (rotation about its main diagonal), $[\mathbf{T}^{t}]_{ij} = [\mathbf{T}]_{ji}$. In fact, this is simple for a 2×2 real symmetric ($M_{12} = M_{21}$) matrix:

$$\mathbf{M} = \begin{bmatrix} M_{11} & M_{12} \\ M_{12} & M_{22} \end{bmatrix}$$

and the real orthogonal transformation matrix $\mathbf{T}$ is simple to obtain:

$$\mathbf{T} = \begin{bmatrix} \cos\theta & -\sin\theta \\ \sin\theta & \cos\theta \end{bmatrix} \tag{5.17a}$$

where angle θ is obtained from

$$\tan2\theta = \frac{2M_{12}}{M_{11} - M_{22}} \tag{5.17b}$$

EXAMPLE

$$\mathbf{M} = \begin{bmatrix} 5 & 2 \\ 2 & 3 \end{bmatrix}$$

$$\tan2\theta = \frac{4}{5 - 3}$$

so that $\theta = 31.72°$. So

$$\mathbf{T} = \begin{bmatrix} 0.85 & -0.53 \\ 0.53 & 0.85 \end{bmatrix}$$

Note that the *columns* of $\mathbf{T}$ can be thought of as vectors in two-dimensional space and, by this procedure, are automatically normalized to a length (in space) of unity (i.e., $\sqrt{\cos^2\theta + \sin^2\theta} = 1$). They are also perpendicular to each other in space. Hence it is said to be an *orthogonal transformation.* You can check that this $\mathbf{T}$ has the properties

$$\mathbf{T}^{-1} = \mathbf{T}^t$$

$$\mathbf{T}^{-1}\mathbf{M}\mathbf{T} = \mathbf{T}^t\mathbf{M}\mathbf{T} = \begin{bmatrix} 6.24 & 0 \\ 0 & 1.76 \end{bmatrix}$$

Case 1: Homogeneous Media, $\mathbf{L} = (1/v^2)\mathbf{C}^{-1}$ First diagonalize $\mathbf{L}$ with the method above to give

$$\mathbf{T}^{-1}\mathbf{L}\mathbf{T} = \mathbf{T}^t\mathbf{L}\mathbf{T} = \begin{bmatrix} l_{mG} & 0 \\ 0 & l_{mR} \end{bmatrix}$$

But $\mathbf{L} = (1/v^2)\mathbf{C}^{-1}$, where the medium is homogeneous and

$$v = \frac{1}{\sqrt{\mu\varepsilon}}$$

Hence

$$\mathbf{T}^{-1}\mathbf{L}\mathbf{T} = \mathbf{T}^t\mathbf{L}\mathbf{T} = \begin{bmatrix} l_{mG} & 0 \\ 0 & l_{mR} \end{bmatrix}$$

$$= \frac{1}{v^2}\mathbf{T}^{-1}\mathbf{C}^{-1}\mathbf{T} = \frac{1}{v^2}\mathbf{T}^t\mathbf{C}^{-1}\mathbf{T}$$

Therefore,

$$\mathbf{T}^{-1}\mathbf{C}\mathbf{T} = \mathbf{T}^t\mathbf{C}\mathbf{T} = \begin{bmatrix} \dfrac{1}{v^2 l_{mG}} & 0 \\ 0 & \dfrac{1}{v^2 l_{mR}} \end{bmatrix}$$

$$= \begin{bmatrix} c_{mG} & 0 \\ 0 & c_{mR} \end{bmatrix}$$

So we can define

$$\boxed{\begin{aligned} \mathbf{T}_V &= \mathbf{T} \\ \mathbf{T}_I &= \mathbf{T} \end{aligned}} \tag{5.18}$$

The individual mode characteristic impedances are

$$\boxed{\begin{aligned} Z_{CmG} &= v l_{mG} \\ Z_{CmR} &= v l_{mR} \end{aligned}} \tag{5.19}$$

and for this *homogeneous medium*, all mode velocities of propagation are equal:

$$\boxed{v_{mG} = v_{mR} = v = \frac{1}{\sqrt{\mu\varepsilon}}} \tag{5.20}$$

Case 2: Inhomogeneous Media, $\mathbf{L} \neq (1/v^2)\mathbf{C}^{-1}$ First diagonalize $\mathbf{C}$ with an orthogonal transformation:

$$\mathbf{U}^{-1}\mathbf{C}\mathbf{U} = \mathbf{U}^t\mathbf{C}\mathbf{U} = \boldsymbol{\theta}^2 = \begin{bmatrix} \theta_1^2 & 0 \\ 0 & \theta_2^2 \end{bmatrix}$$

Since C is positive definite, all its eigenvalues are positive and nonzero, so its square roots will be real and nonzero. Then form

$$\theta U^t L U \theta$$

But this is real and symmetric, so find another orthogonal transformation S to diagonalize this:

$$S^t(\theta U^t L U \theta)S = \begin{bmatrix} \Lambda_1^2 & 0 \\ 0 & \Lambda_2^2 \end{bmatrix}$$

Then define a 2×2 matrix T as

$$T = U\theta S$$

Normalize the columns of T to a length of unity:

$$\begin{aligned} T_{\text{norm}} &= U\theta S\alpha \\ &= T\alpha \end{aligned}$$

where α is a 2×2 diagonal matrix with main diagonal entries

$$\alpha_{11} = \frac{1}{\sqrt{T_{11}^2 + T_{21}^2}}$$
$$\alpha_{22} = \frac{1}{\sqrt{T_{12}^2 + T_{22}^2}}$$

Then define the mode transformations as

$$\boxed{\begin{aligned} T_I &= T_{\text{norm}} = U\theta S\alpha \\ T_V &= U\theta^{-1}S\alpha^{-1} \end{aligned}} \tag{5.21}$$

You can check that

$$\begin{aligned} T_V^{-1}LT_I &= \alpha S^t \theta U^t L U \theta S \alpha \\ &= \begin{bmatrix} \alpha_{11}^2 \Lambda_1^2 & 0 \\ 0 & \alpha_{22}^2 \Lambda_2^2 \end{bmatrix} \end{aligned}$$

and

$$\begin{aligned} T_I^{-1}CT_V &= \alpha^{-1}S^t\theta^{-1}U^tCU\theta^{-1}S\alpha^{-1} \\ &= \begin{bmatrix} \dfrac{1}{\alpha_{11}^2} & 0 \\ 0 & \dfrac{1}{\alpha_{22}^2} \end{bmatrix} \end{aligned}$$

So

$$l_{mG} = \alpha_{11}^2 \Lambda_1^2$$
$$l_{mR} = \alpha_{22}^2 \Lambda_2^2$$

and

$$c_{mG} = \frac{1}{\alpha_{11}^2}$$
$$c_{mR} = \frac{1}{\alpha_{22}^2}$$

Hence the characteristic impedances of the modes are

$$Z_{CmG} = \sqrt{\frac{l_{mG}}{c_{mG}}} = \alpha_{11}^2 \Lambda_1$$

$$Z_{CmR} = \sqrt{\frac{l_{mR}}{c_{mR}}} = \alpha_{22}^2 \Lambda_2$$

(5.22)

and the individual mode velocities of propagation are

$$v_{mG} = \frac{1}{\sqrt{l_{mG}c_{mG}}} = \frac{1}{\Lambda_1}$$

$$v_{mR} = \frac{1}{\sqrt{l_{mR}c_{mR}}} = \frac{1}{\Lambda_2}$$

(5.23)

The previous development allows us to develop a SPICE subcircuit model for direct insertion into a SPICE analysis to easily determine the crosstalk for any *lossless* multiconductor transmission line (MTL) problem.

The SPICEMTL Computer Program (SPICEMTL.FOR, SPICEMTL. EXE) SPICEMTL.FOR is a FORTRAN program for generating a SPICE subcircuit model of a *lossless* MTL. The program SPICEMTL.EXE is the compiled and executable file. SPICEMTL.EXE reads line dimensions from the file SPICEMTL.IN and then reads the **L** and **C** matrices for the line from PUL.DAT. It generates a subcircuit model of the line that is output to the file SPICEMTL.OUT. The subcircuit model shown in Fig. 5.2 is delivered to SPICEMTL.OUT and starts with the usual subcircuit line,

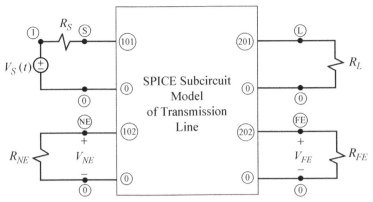

FIGURE 5.3. Connection of the nodes of the subcircuit MTL to the external nodes of the actual circuit.

```
.SUBCKT MTL 101 102 201 202
```

and ends with

```
.ENDS MTL
```

The internal nodes of the subcircuit model are connected to the nodes of the external transmission-line circuit as shown in Fig. 5.3. In Fig. 5.3, the nodes of the external MTL are named S, L, NE, and FE for bookkeeping purposes. It is very important that the user connect the proper nodes in the subcircuit model to the corresponding conductor numbers in PUL.DAT computed for this structure. Here the entries in PUL.DAT were computed for $1 = G$ and $2 = R$. Hence nodes 101 and 201 are associated with the ends of the generator conductor, and nodes 102 and 202 are associated with the ends of the receptor conductor. The connections are made in the SPICE program with the statement

```
XMTL S NE L FE MTL
```

This establishes the connection $S \rightarrow 101$, $NE \rightarrow 102$, $L \rightarrow 201$, and $FE \rightarrow 202$.

EXAMPLE

Consider the example shown in Fig. 5.4. The cross-sectional configuration is a PCB having land widths of 15 mils, edge-to-edge separations of 45 mils, and a

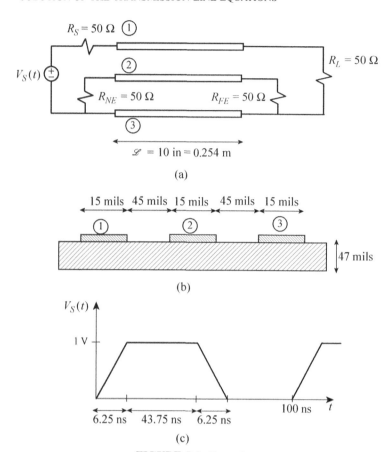

FIGURE 5.4. Example.

board thickness of 47 mils. The board is constructed of glass-epoxy having $\varepsilon_r \cong 4.7$, and the lands are of total length 10 in $= 0.254$ m. The lands are terminated in 50-Ω resistors, and the source voltage $V_S(t)$ is a 10-MHz clock having a 1-V amplitude, a 50% duty cycle, and 6.25 ns 0 to 100% rise and fall times, which are equivalent to $\tau_{r,10-90\%} = \tau_{f,10-90\%} = 5$ ns.

The PCB.IN data are

```
3        =TOTAL NUMBER OF LANDS
30.      =NUMBER OF CONDUCTOR SUBSECTIONS
3        =NUMBER OF REFERENCE LAND
15.0     =LAND WIDTH (mils)
45.0     =EDGE-TO-EDGE LAND SEPARATION (mils)
```

```
47.0   =BOARD THICKNESS (mils)
4.7    =BOARD RELATIVE DIELECTRIC CONSTANT
```

The output data for L and C are in PUL.DAT:

```
1     1        1.38314E-06     =L(1, 1)
1     2        6.91570E-07     =L(1, 2)
2     2        1.10706E-06     =L(2, 2)
1     1        2.96949E-11     =C(1, 1)
1     2       -2.02619E-11     =C(1, 2)
2     2        4.05238E-11     =C(2, 2)
1     1        1.16983E-11     =C0(1, 1)
1     2       -7.30777E-12     =C0(1, 2)
2     2        1.46155E-11     =C0(2, 2)
```

```
NUMBER OF LANDS= 3
NUMBER OF DIVISIONS PER LAND= 30
REFERENCE LAND= 3
LAND WIDTH (mils)= 1.500E+01
EDGE-TO-EDGE SEPARATION (mils)= 4.500E+01
BOARD THICKNESS (mils)= 4.700E+01
RELATIVE DIELECTRIC CONSTANT= 4.700E+00
```

The SPICEMTL.IN data are

```
3       =Total Number of Conductors (N+1)
0.254   =Total Line Length (m)
```

The SPICEMTL.OUT file contains the PSPICE subcircuit model generated:

```
*SUBCIRCUIT MODEL OF A MULTICONDUCTOR TRANSMISSION
LINE*
* NUMBER OF CONDUCTORS= 3
* TOTAL LINE LENGTH (METERS)= 2.54000E-01
*    L(1, 1)= 1.38314E-06
*    L(1, 2)= 6.91570E-07
*    L(2, 2)= 1.10706E-06
*    C(1, 1)= 2.96949E-11
*    C(1, 2)=-2.02619E-11
*    C(2, 2)= 4.05238E-11
```

```
.SUBCKT MTL
+ 101
+ 102
+ 201
+ 202
V101 101 301
EC101 301 0 POLY(2)
+ (501,0)
+ (502,0)
+ 0
+ 1.000000E+00
+ -1.046409E-07
FC101 0 501 POLY(2)
+ V101
+ V102
+ 0
+ 1.000000E+00
+ 5.000001E-01
V201 201 401
EC201 401 0 POLY(2)
+ (601,0)
+ (602,0)
+ 0
+ 1.000000E+00
+ -1.046409E-07
FC201 0 601 POLY(2)
+ V201
+ V202
+ 0
+ 1.000000E+00
+ 5.000001E-01
V102 102 302
EC102 302 0 POLY(2)
+ (501,0)
+ (502,0)
+ 0
+ 5.000001E-01
+ 1.118034E+00
FC102 0 502 POLY(2)
+ V101
+ V102
+ 0
```

```
+ -1.046409E-07
+ 1.118034E+00
V202 202 402
EC202 402 0 POLY(2)
+ (601,0)
+ (602,0)
+ 0
+ 5.000001E-01
+ 1.118034E+00
FC202 0 602 POLY(2)
+ V201
+ V202
+ 0
+ -1.046409E-07
+ 1.118034E+00
T101 501 0 601 0 Z0= 2.658917E+02 TD= 1.321281E-09
T102 502 0 602 0 Z0= 1.096493E+02 TD= 1.410781E-09
.ENDS MTL
```

Adding the terminal circuitry to this subcircuit model gives the SPICE.IN file:

```
SPICE MTL MODEL; INHOMOGENEOUS MEDIUM
VS 1 0 PULSE(0 1 0 6.25N 6.25N 43.75N 100N)
*VS 1 0 AC 1 0
RS 1 S 50
RL L 0 50
RNE NE 0 50
RFE FE 0 50
XMTL S NE L FE MTL
.TRAN 0.1N 100N 0 0.1N
*.AC DEC 50 10K 1000MEG
.PROBE
*SUBCIRCUIT MODEL OF A MULTICONDUCTOR TRANSMISSION
LINE*
* NUMBER OF CONDUCTORS= 3
* TOTAL LINE LENGTH (METERS)= 2.54000E-01
*    L(1, 1)= 1.38314E-06
*    L(1, 2)= 6.91570E-07
*    L(2, 2)= 1.10706E-06
*    C(1, 1)= 2.96949E-11
*    C(1, 2)=-2.02619E-11
```

```
*   C(2, 2)= 4.05238E-11
.SUBCKT MTL
+ 101
+ 102
+ 201
+ 202
V101 101 301
EC101 301 0 POLY(2)
+ (501,0)
+ (502,0)
+ 0
+ 1.000000E+00
+ -1.046409E-07
FC101 0 501 POLY(2)
+ V101
+ V102
+ 0
+ 1.000000E+00
+ 5.000001E-01
V201 201 401
EC201 401 0 POLY(2)
+ (601,0)
+ (602,0)
+ 0
+ 1.000000E+00
+ -1.046409E-07
FC201 0 601 POLY(2)
+ V201
+ V202
+ 0
+ 1.000000E+00
+ 5.000001E-01
V102 102 302
EC102 302 0 POLY(2)
+ (501,0)
+ (502,0)
+ 0
+ 5.000001E-01
+ 1.118034E+00
FC102 0 502 POLY(2)
+ V101
+ V102
```

```
+ 0
+ -1.046409E-07
+  1.118034E+00
V202 202 402
EC202 402 0 POLY(2)
+ (601,0)
+ (602,0)
+ 0
+ 5.000001E-01
+ 1.118034E+00
FC202 0 602 POLY(2)
+ V201
+ V202
+ 0
+ -1.046409E-07
+  1.118034E+00
T101 501 0 601 0 Z0= 2.658917E+02 TD= 1.321281E-09
T102 502 0 602 0 Z0= 1.096493E+02 TD= 1.410781E-09
.ENDS MTL
.END
```

Note that there are two distinct mode velocities: $v_{m1} = \mathcal{L}/T_{D1}$ $= 1.922 \times 10^8$ m/s and $v_{m2} = \mathcal{L}/T_{D2} = 1.8 \times 10^8$ m/s. Both of these are approximately 60% of the speed of light in free space.

Running the SPICE.IN program determines the near- and far-end crosstalk voltages as shown in Fig. 5.5. These crosstalk voltages are verified, *experimentally*, in:

C. R. Paul, *Introduction to Electromagnetic Compatibility*, 2nd ed., Wiley-Interscience, Hoboken, NJ, 2006.

C. R. Paul, *Analysis of Multiconductor Transmission Lines*, 2nd ed., Wiley-Interscience, Hoboken, NJ, 2008.

The near- and far-end frequency-domain transfer functions $\hat{V}_{NE}/\hat{V}_S$ and $\hat{V}_{FE}/\hat{V}_S$ can be obtained by adding

```
VS 1 0 AC 1 0
.AC DEC 50 10K 1000MEG
.PRINT AC VM(NE) VP(NE) VM(FE) VP(FE)
```

giving the magnitudes of the frequency-domain transfer functions shown in Figure 5.6.

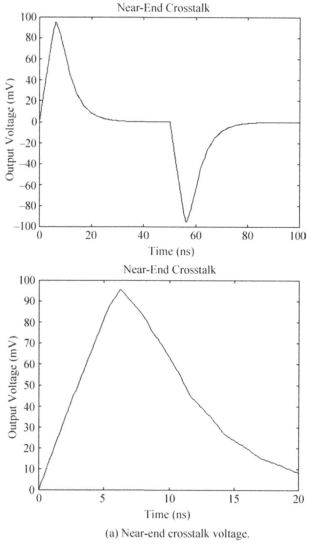

(a) Near-end crosstalk voltage.

FIGURE 5.5. (a) Near- and (b) far-end crosstalk voltages for the problem of Fig. 5.4.

The line (in free space with the dielectric substrate removed) is one wavelength at

$$f = \frac{v_0}{\mathscr{L}} = \frac{3 \times 10^8}{0.254} = 1.18 \, \text{GHz}$$

However, because of the inhomogeneous medium, there are two mode velocities: $v_1 = 1.92 \times 10^8$ and $v_2 = 1.8 \times 10^8$. From the plots of the transfer functions the frequencies where the line appears to be one wavelength are around 700 MHz. The inverse of the two mode-time delays generated by

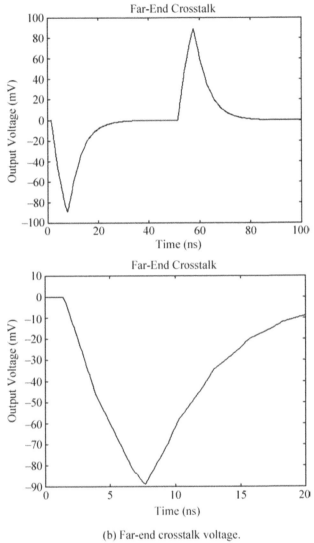

(b) Far-end crosstalk voltage.

FIGURE 5.5. (*Continued*)

the SPICE subcircuit model give these as $1/T_{D1} = 756\,\text{MHz}$ and $1/T_{D2} = 708\text{MHz}$. Notice that the magnitudes of the crosstalk transfer functions increase linearly with frequency for frequencies where the line is electrically short. From the frequency-response plots in Fig. 5.6, this appears to be below 70 MHz. This linear increase of the transfer function with frequency where the line is electrically short is a general result that we find in the final section of this Chapter.

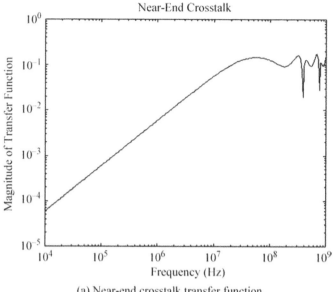

(a) Near-end crosstalk transfer function.

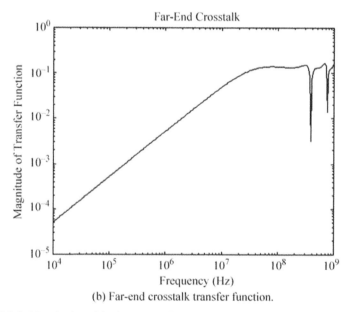

(b) Far-end crosstalk transfer function.

FIGURE 5.6. Magnitudes of the frequency-domain transfer functions for the problem shown in Fig. 5.4, for (a) near-end and (b) far-end crosstalk.

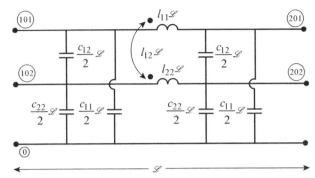

FIGURE 5.7. Lumped-pi approximate model for the three-conductor line.

5.3   LUMPED-CIRCUIT APPROXIMATE MODELS OF THE LINE

As was the case for two-conductor lines, we can construct lumped equivalent circuits of the line for frequencies of the source where the line is *electrically short*. The most common type is the lumped-pi circuit shown in Fig. 5.7, wherein the per-unit-length self- and mutual inductances are multiplied by the line length, $\mathscr{L}$, and placed in the middle, and the per-unit-length self- and mutual capacitances are multiplied by the line length, $\mathscr{L}$, then split and placed on both ends. This is done to make the model symmetrical, as is the actual line.

Lumped-Circuit Approximate Models of the Line (SPICELPI.FOR, SPICELPI.EXE)   The FORTRAN program SPICELPI.FOR constructs a subcircuit model for a one-section lumped-pi equivalent circuit as shown in Fig. 5.7. The executable file SPICELPI.EXE reads the line dimensions from SPICELPI.IN and the line per-unit-length **L** and **C** matrices from PUL.DAT and constructs the subcircuit model, which is output to the file SPICELPI. OUT. The source and load circuitry are added to this to give a complete SPICE circuit.

For the example shown in Fig. 5.4, the SPICELPI.IN file is

```
3       =Total Number of Conductors (N+1)
0.254   =Total Line Length (m)
```

The SPICE subcircuit model generated is output to the file SPICELPI. OUT:

```
*SUBCIRCUIT MODEL OF A MULTICONDUCTOR TRANSMISSION
LINE*
*LUMPED-PI APPROXIMATE STRUCTURE*
```

```
*      TOTAL NUMBER OF CONDUCTORS= 3
*      TOTAL LINE LENGTH (METERS)= 2.54000E-01
*      L(1, 1)= 1.38314E-06
*      L(1, 2)= 6.91570E-07
*      L(2, 2)= 1.10706E-06
*      C(1, 1)= 2.96949E-11
*      C(1, 2)=-2.02619E-11
*      C(2, 2)= 4.05238E-11
.SUBCKT LPI
+ 101
+ 102
+ 201
+ 202
L101   101   201   3.51318E-07
C101   101   0   1.19799E-12
C201   201   0   1.19799E-12
L102   102   202   2.81193E-07
C102   102   0   2.57326E-12
C202   202   0   2.57326E-12
K101   L101   L102   5.58879E-01
CM101   101   102   2.57326E-12
CM201   201   202   2.57326E-12
.ENDS LPI
```

The terminations are added to this to give the SPICE.IN complete file:

```
SPICE LUMPED PI MODEL
VS 1 0 PULSE(0 1 0 6.25N 6.25N 43.75N 100N)
*VS 1 0 AC 1 0
RS 1 S 50
RL L 0 50
RNE NE 0 50
RFE FE 0 50
XLPI S NE L FE LPI
.TRAN 0.1 100N 0 0.1N
*.AC DEC 10K 1000MEG
.PROBE
*SUBCIRCUIT MODEL OF A MULTICONDUCTOR TRANSMISSION
LINE*
```

```
*LUMPED-PI APPROXIMATE STRUCTURE*
* TOTAL NUMBER OF CONDUCTORS= 3
* TOTAL LINE LENGTH (METERS)= 2.54000E-01
* L(1, 1)= 1.38314E-06
* L(1, 2)= 6.91570E-07
* L(2, 2)= 1.10706E-06
* C(1, 1)= 2.96949E-11
* C(1, 2)=-2.02619E-11
* C(2, 2)= 4.05238E-11
.SUBCKT LPI
+ 101
+ 102
+ 201
+ 202
L101 101 201 3.51318E-07
C101 101 0 1.19799E-12
C201 201 0 1.19799E-12
L102 102 202 2.81193E-07
C102 102 0 2.57326E-12
C202 202 0 2.57326E-12
K101 L101 L102 5.58879E-01
CM101 101 102 2.57326E-12
CM201 201 202 2.57326E-12
.ENDS LPI
.END
```

Predictions of the near- and far-end crosstalk voltages from the MTL subcircuit model and from the lumped-pi subcircuit model are compared in Fig. 5.8. Predictions of the magnitudes of the frequency-domain transfer functions computed by the MTL subcircuit and lumped-pi models are compared in Fig. 5.9.

The line is electrically short at around 70 MHz. The bandwidth of the source voltage waveform is approximately

$$\text{BW} = \frac{1}{\tau_r} = \frac{1}{6.25 \text{ ns}} = 160 \text{ MHz}$$

Hence it is not surprising that the lumped-pi model gives a reasonable prediction of the exact MTL model results.

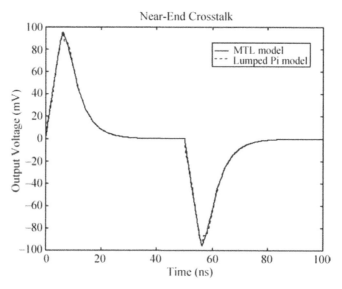

(a) Near-end crosstalk voltage predictions.

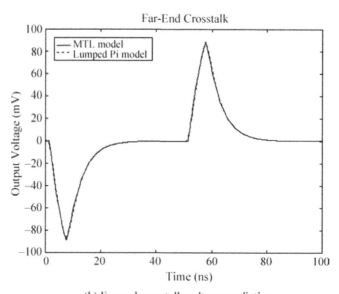

(b) Far-end crosstalk voltage predictions.

FIGURE 5.8. Comparison of the predictions of the the (a) near- and (b) far-end crosstalk voltages of the SPICE subcircuit model and the lumped-pi subcircuit model for the problem of Fig. 5.4.

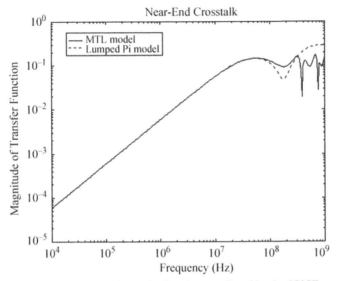

(a) Near-end transfer functions predicted by the SPICE
and Lumped-Pi subcircuit models.

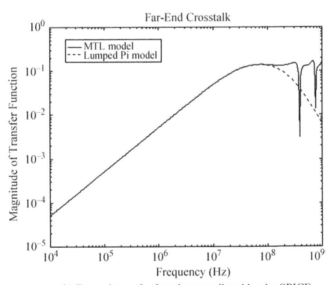

(b) Far-end transfer functions predicted by the SPICE
and Lumped-Pi subcircuit models.

FIGURE 5.9. Comparison of the (a) near- and (b) far-end transfer functions predicted
by the SPICE subcircuit models and the lumped-pi subcircuit models for the problem of
Fig. 5.4.

5.4   THE INDUCTIVE–CAPACITIVE COUPLING APPROXIMATE MODEL

Finally, we obtain a very simple but approximate crosstalk model. There are two restrictions on its use:

1. *The lines must be weakly coupled.* The voltage and current in the driven or generator circuit, V_G and I_G, create electric and magnetic fields that couple to the receptor circuit and induce in it voltage V_R and current I_R. In turn, the voltage V_R and current I_R also generate electric and magnetic fields that couple to the generator circuit, producing additional voltages and currents that add to V_G and I_G, and so on. The lines are said to be *weakly coupled* if the voltage and current induced in the receptor circuit, V_R and I_R, are much smaller than V_G and I_G, so that the additional voltage and current contributions induced back into the generator circuit from the receptor circuit are insignificant. Hence for weakly coupled lines this induction is a one-way process: from the generator circuit to the receptor circuit. We can quantitatively judge whether this is satisfied by examining the inductive coupling coefficient between the two lines:

$$k = \frac{l_m}{\sqrt{l_G l_R}} \tag{5.24}$$

This is like the coupling coefficient between the primary and the secondary of a transformer. If $k \ll 1$, we say that the circuits are weakly coupled.

2. *The second criterion is that the line must be electrically very short at the largest significant frequency of $V_S(t)$ (its bandwidth).* Essentially, this means that the voltage and current of the generator line are essentially the same at all points along the generator line, so that we can say approximately that

$$V_G(z, t) \cong \frac{R_L}{R_S + R_L} V_S(t) \quad \text{for all } 0 \le z \le \mathscr{L}$$

$$I_G(z, t) \cong \frac{V_S(t)}{R_S + R_L} \qquad \text{for all } 0 \le z \le \mathscr{L}$$

Hence the voltage and current of the generator line are as though they were dc.

Hence we may approximately represent the coupled lines as shown in Fig. 5.10, and the total *mutual* inductance and *mutual* capacitance are denoted as $L_m = l_m \mathscr{L}$ and $C_m = c_m \mathscr{L}$. From this approximate circuit, the crosstalk

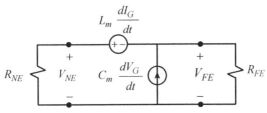

FIGURE 5.10. Weakly coupled, electrically short crosstalk prediction approximate model.

voltages are obtained, by superposition, as

$$V_{NE}(t) \cong \frac{R_{NE}}{R_{NE}+R_{FE}} L_m \frac{dI_G}{dt} + \frac{R_{NE}R_{FE}}{R_{NE}+R_{FE}} C_m \frac{dV_G}{dt} \qquad (5.25a)$$

$$V_{FE}(t) \cong -\frac{R_{FE}}{R_{NE}+R_{FE}} L_m \frac{dI_G}{dt} + \frac{R_{NE}R_{FE}}{R_{NE}+R_{FE}} C_m \frac{dV_G}{dt} \qquad (5.25b)$$

Substituting the approximate relations for V_G and I_G gives

$$V_{NE}(t) \cong \left(\frac{R_{NE}}{R_{NE}+R_{FE}} L_m + \frac{R_{NE}R_{FE}}{R_{NE}+R_{FE}} C_m R_L \right) \frac{1}{R_S+R_L} \frac{dV_S(t)}{dt} \qquad (5.26a)$$

$$V_{FE}(t) \cong \left(-\frac{R_{FE}}{R_{NE}+R_{FE}} L_m + \frac{R_{NE}R_{FE}}{R_{NE}+R_{FE}} C_m R_L \right) \frac{1}{R_S+R_L} \frac{dV_S(t)}{dt}$$

$$(5.26b)$$

This shows that the near- and far-end crosstalk voltages are proportional to the derivative of $V_S(t)$! Hence, if the source voltage $V_S(t)$ is a trapezoidal-shaped digital waveform having rise time τ_r and fall time τ_f and the restrictions on this model apply, the near- and far-end time-domain crosstalk voltage waveforms should appear to be rectangular pulses occurring during the rise and fall times of the pulse train, and the heights of the pulses should be proportional to the *slew rates* of $V_S(t)$: A/τ_r and A/τ_f.

The frequency-domain, phasor crosstalk voltages are obtained by substituting $j\omega$ for all time derivatives:

$$\hat{V}_{NE} \cong j\omega\left(\frac{R_{NE}}{R_{NE}+R_{FE}}L_m + \frac{R_{NE}R_{FE}}{R_{NE}+R_{FE}}C_mR_L\right)\frac{1}{R_S+R_L}\hat{V}_S \qquad (5.27a)$$

$$\hat{V}_{FE}(t) \cong j\omega\left(-\frac{R_{FE}}{R_{NE}+R_{FE}}L_m + \frac{R_{NE}R_{FE}}{R_{NE}+R_{FE}}C_mR_L\right)\frac{1}{R_S+R_L}\hat{V}_S$$

$$(5.27b)$$

This shows that the frequency-response crosstalk voltage transfer functions, $\hat{V}_{NE}/\hat{V}_S$ and $\hat{V}_{FE}/\hat{V}_S$, increase linearly with frequency and have a phase angle of $\pm 90°$.

This model is said to be the *inductive–capacitive coupling model* since the crosstalk is produced by the total mutual inductance between the two circuits, L_m, and the total mutual capacitance between the two circuits, C_m. Equations (5.27) show that for the frequencies where the line is electrically short, the frequency-domain crosstalk voltage transfer functions must increase linearly with frequency. Hence we can obtain a criterion for when the line is electrically very short for a particular time-domain source voltage waveform, $V_S(t)$, and when this model is valid. If $V_S(t)$ represents a digital clock waveform with equal rise and fall times, the bandwidth is approximately $\text{BW} \cong 1/\tau_r$. For the line to be electrically very short at this maximum frequency, we must have

$$\mathscr{L} \ll \frac{1}{10}\frac{v}{\text{BW}}$$

or

$$\boxed{\tau_r \gg 10T_D} \qquad (5.28)$$

where the time delay, T_D, is the maximum of the mode time delays of the line. But for practical lines they do not differ by much, so this choice isn't crucial.

EXAMPLE

Consider the example shown in Fig. 5.4. From the SPICEMTL computations, the one-way time delays of the modes are $T_{D1} = 1.32$ ns and $T_{D2} = 1.41$ ns. Hence for the lines to be electrically very short, the rise time should be greater than perhaps 100 ns. So we will arbitrarily reset the rise and fall times of the pulse to 100 ns and the period to 500 ns (2 MHz). The coupling coefficient is $k = l_m/\sqrt{l_G l_R} = 0.559$. This is bordering on the boundary between weakly coupled and strongly coupled. The exact time-domain near- and far-end crosstalk voltages computed with the SPICE subcircuit model generated by SPICEMTL.FOR are shown in Fig. 5.11.

Observe that the crosstalk voltages appear, as expected, as rectangles during the rise and fall times of $V_S(t)$, which are $0 \leq t \leq 100$ ns and $250 \leq t \leq 350$ ns. From the PUL.DAT data, $l_m = 6.9157 \times 10^{-7}$ and $c_m = 2.02619 \times 10^{-11}$. Since the total line length is $\mathcal{L} = 0.254$ m, the total mutual inductance and capacitance are $L_m = l_m \mathcal{L} = 1.757 \times 10^{-7}$ H and $C_m = c_m \mathcal{L} = 5.147 \times 10^{-12}$ F. The derivative (slew rate) of $V_S(t)$ is

$$\frac{dV_S(t)}{dt} = \frac{1 \text{V}}{100 \text{ ns}} = 10^7 \, {}^V\!/_S$$

Hence we compute the near- and far-end crosstalk voltages [the heights of the pulses occurring during the rise and fall times of $V_S(t)$] from (5.26) as

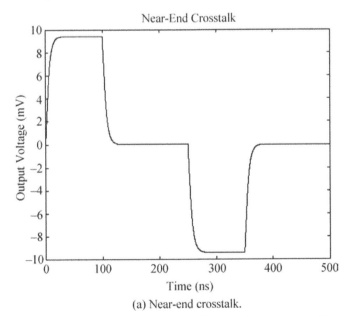

(a) Near-end crosstalk.

FIGURE 5.11. (a) Near-end and (b) far-end crosstalk predictions for the problem of Fig. 5.4.

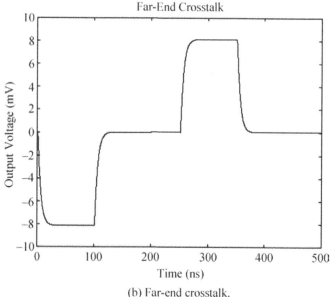

(b) Far-end crosstalk.

FIGURE 5.11. (*Continued*)

$$V_{NE}(t) \cong \left(\frac{R_{NE}}{R_{NE} + R_{FE}} L_m + \frac{R_{NE}R_{FE}}{R_{NE} + R_{FE}} C_m R_L \right) \frac{1}{R_S + R_L} \frac{dV_S(t)}{dt}$$

$$= 9.426 \, \text{mV}$$

$$V_{FE}(t) \cong \left(-\frac{R_{FE}}{R_{NE} + R_{FE}} L_m + \frac{R_{NE}R_{FE}}{R_{NE} + R_{FE}} C_m R_L \right) \frac{1}{R_S + R_L} \frac{dV_S(t)}{dt}$$

$$= -8.14 \, \text{mV}$$

The *exact* voltages computed using PSPICE and the SPICEMTL subcircuit model and shown in Fig. 5.11 are $V_{NE} = 9.4263$ mV and $V_{FE} = -8.1396$ mV!

PROBLEMS

5.1 Determine the 2×2 transformation matrix that diagonalizes
$\mathbf{M} = \begin{bmatrix} 7 & 3 \\ 3 & 5 \end{bmatrix}$ as $\mathbf{T}^{-1}\mathbf{M}\mathbf{T} = \mathbf{T}^t\mathbf{M}\mathbf{T} = \begin{bmatrix} \Lambda_1 & 0 \\ 0 & \Lambda_2 \end{bmatrix}$ and determine Λ_1
and Λ_2. $[\mathbf{T} = \begin{bmatrix} 0.811 & -0.585 \\ 0.585 & 0.811 \end{bmatrix}, \Lambda_1 = 9.162, \Lambda_2 = 2.838]$

5.2   Use SPICEMTL.FOR to compute the near- and far-end crosstalk voltage waveforms for the situation shown in Fig. P5.2.

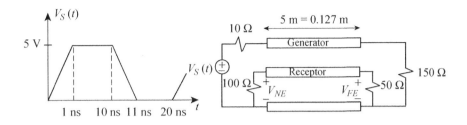

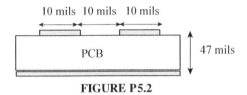

FIGURE P5.2

5.3   For the situation of Problem 5.2, change the rise and fall times to a value that satisfies (5.28) and recompute the near- and far-end voltage waveforms using **(a)** SPICEMTL.FOR and **(b)** SPICELPI.FOR. Compute the peak near- and far-end crosstalk voltages using the simple inductive-coupling, capacitive-coupling coupling mode, and compare the results.

5.4   Use SPICEMTL.FOR to compute the near- and far-end crosstalk voltage waveforms for the situation shown in Fig. P5.4.

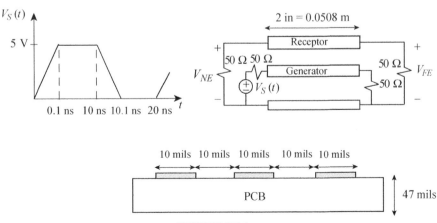

FIGURE P5.4

5.5 For the situation of Problem 5.4, change the rise and fall times to a value that satisfies (5.28), and recompute the near- and far-end voltage waveforms using (**a**) SPICEMTL.FOR and (**b**) SPICELPI.FOR. Compute the peak near- and far-end crosstalk voltages using the simple inductive-coupling, capacitive-coupling coupling mode, and compare the results.

6

SOLUTION OF THE TRANSMISSION-LINE EQUATIONS FOR THREE-CONDUCTOR LOSSY LINES

In this final chapter we determine the effect of line losses on the near- and far-end crosstalk voltages, $V_{NE}(t)$ and $V_{FE}(t)$. There are two sources of the line loss:

1. The resistance of the line conductors
2. Losses in the surrounding dielectric medium, as in the case of the dielectric within a coaxial cable or the dielectric board of a PCB

We will find that the resistance of the line conductors begin to increase from their constant dc values to a rate of $\sqrt{f}$ above a certain frequency where the conductor cross-sectional dimensions begin to become on the order of a few *skin depths*:

$$\delta = \frac{1}{\sqrt{\pi f \mu \sigma}} \tag{6.1}$$

where f is the frequency of the currents on them, μ is the permeability of the conductors (which is generally assumed to be nonmagnetic and is therefore

Transmission Lines in Digital and Analog Electronic Systems: Signal Integrity and Crosstalk, By Clayton R. Paul
Copyright © 2010 John Wiley & Sons, Inc.

that of free space, $\mu_0 = 4\pi \times 10^{-7}$), and σ is the conductivity of copper, of which most conductors are constructed: $\sigma_{Cu} = 5.8 \times 10^7$ S/m. The dielectric loss is also a function of the frequency of the source and is significant only over certain frequency intervals. As the frequencies of the clocks and data streams continue to move into the gigahertz range, these losses continue to increase and are beginning to cause signal integrity problems. We will find a very simple-to-understand problem that they cause: level shifts in the crosstalk waveforms that we refer to as *common-impedance coupling*. This is a simple Ohm's law type of direct coupling that occurs in addition to the electric and magnetic field coupling between the generator and receptor circuits. This can also cause logic errors and other signal integrity problems and is relatively easy to understand. The losses will also have the effect of "smoothing" the abrupt time-domain transitions we have seen in the lossless-line case that are due to reflections at the terminations.

Since the PSPICE program contains only transmission-line models of *lossless lines*, it cannot be used to predict these effects of the line losses either in signal integrity for two-conductor lines or through the SPICEMTL subcircuit model generated by the SPICEMTL.FOR program to predict their effects on crosstalk for three-conductor lines. The CD that accompanies this book contains a program MTL.FOR and its compiled and executable version, MTL.EXE, which can be used to model the effects of these frequency-dependent line losses in the frequency domain and then predict these frequency-domain effects indirectly in the time domain by using superposition, although this procedure assumes linear terminations.

6.1 THE TRANSMISSION-LINE EQUATIONS FOR THREE-CONDUCTOR LOSSY LINES

A per-unit-length model of a two-conductor *lossy* line is shown in Fig. 6.1. The effect of the line conductors is modeled by the per-unit-length resistance $r(f)$, which is generally frequency dependent. The effect of the losses in the surrounding dielectric medium (which may be inhomogeneous) is modeled by the per-unit-length conductance $g(f)$, which is also, in general, frequency dependent. These frequency-dependent parameters, $r(f)$ and $g(f)$, are easily included in the *phasor* transmission-line equations in the frequency domain, but their inclusion directly in a time-domain analysis is obviously problematic: How do we model directly parameters that are functions of frequency in the time domain? The phasor (frequency-domain) transmission-line equations, which include the effects of these two loss mechanisms, including their frequency dependence, are easily obtained from the per-unit-length

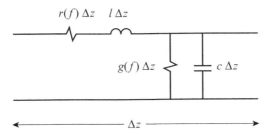

FIGURE 6.1. Per-unit-length model of a two-conductor *lossy* line.

phasor equivalent circuit in the usual way from Fig. 6.1 in the limit as $\Delta z \to 0$ as

$$\frac{d\hat{V}(z)}{dz} = - \underbrace{(r(f) + j\omega l)}_{\hat{z}(f)} \hat{I}(z) \tag{6.2a}$$

$$\frac{d\hat{I}(z)}{dz} = - \underbrace{(g(f) + j\omega c)}_{\hat{y}(f)} \hat{V}(z) \tag{6.2b}$$

where the per-unit-length impedance and admittance are

$$\hat{z}(f) = r(f) + j\omega l \qquad \Omega/\text{m} \tag{6.3a}$$

$$\hat{y}(f) = g(f) + j\omega c \qquad \text{S/m} \tag{6.3b}$$

The *uncoupled* second-order phasor equations are obtained in the usual fashion as

$$\frac{d^2\hat{V}(z)}{dz^2} = \hat{z}(f)\hat{y}(f)\hat{V}(z) \tag{6.4a}$$

$$\frac{d^2\hat{I}(z)}{dz^2} = \hat{z}(f)\hat{y}(f)\hat{I}(z) \tag{6.4b}$$

The general solution to these is fairly simple to obtain in the frequency domain as we have done previously:

$$\hat{V}(z) = \hat{V}^+ e^{-\alpha z} e^{-j\beta z} + \hat{V}^- e^{\alpha z} e^{j\beta z} \tag{6.5a}$$

$$\hat{I}(z) = \frac{\hat{V}^+}{\hat{Z}_C} e^{-\alpha z} e^{-j\beta z} - \frac{\hat{V}^-}{\hat{Z}_C} e^{\alpha z} e^{j\beta z} \tag{6.5b}$$

where the characteristic impedance is, in general, complex:

$$\hat{Z}_C = \sqrt{\frac{\hat{z}(f)}{\hat{y}(f)}} \tag{6.6}$$

$$= Z_C \angle \theta_{Z_C}$$

and the attenuation constant $\alpha(f)$ and propagation constant $\beta(f)$ are obtained from the real and imaginary parts of the complex propagation constant:

$$\hat{\gamma}(f) = \sqrt{\hat{z}(f)\hat{y}(f)} \tag{6.7}$$

$$= \alpha(f) + j\beta(f)$$

The time-domain results are obtained as

$$V(z,t) = V^+ e^{-\alpha z}\cos(\omega t - \beta z + \theta^+) + V^- e^{\alpha z}\cos(\omega t + \beta z + \theta^-) \tag{6.8a}$$

$$I(z,t) = \frac{V^+}{Z_C}e^{-\alpha z}\cos(\omega t - \beta z + \theta^+ - \theta_{Z_C}) - \frac{V^-}{Z_C}e^{\alpha z}\cos(\omega t + \beta z + \theta^- - \theta_{Z_C}) \tag{6.8b}$$

and the undetermined constants have a magnitude and phase of $\hat{V}^+ = V^+\angle\theta^+$ and $\hat{V}^- = V^-\angle\theta^-$. In Chapter 3 we examined briefly the effects that the losses have on the time-domain waveforms: The amplitudes are decreased as $e^{\pm\alpha z}$ and the losses affect the velocity of propagation as $v = \omega/\beta(f)$. Note, however, that since $\alpha(f)$ and $\beta(f)$ depend on frequency, their inclusion in the time-domain result presents logical problems. We solved this dilemma in Chapter 3 by using superposition and combining the responses to the Fourier harmonic components in the output to give the time-domain output waveform, although this process requires linear terminations.

The three-conductor line has direct parallels. The per-unit-length equivalent circuit of a *lossy* three-conductor line is shown in Fig. 6.2. In the limit as $\Delta z \to 0$, we obtain the *phasor* multiconductor transmission-line (MTL) equations, which include the losses of the line from that circuit, as

$$\frac{d}{dz}\hat{\mathbf{V}}(z) = -\underbrace{(\mathbf{R}(f) + j\omega\mathbf{L})}_{\hat{\mathbf{Z}}(f)}\hat{\mathbf{I}}(z) \tag{6.9a}$$

$$\frac{d}{dz}\hat{\mathbf{I}}(z) = -\underbrace{(\mathbf{G}(f) + j\omega\mathbf{C})}_{\hat{\mathbf{Y}}(f)}\hat{\mathbf{V}}(z) \tag{6.9b}$$

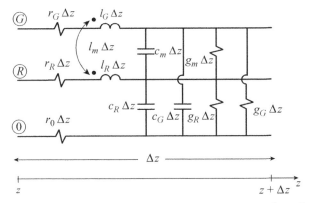

FIGURE 6.2. Per-unit-length model of a three-conductor *lossy* line.

where the 2×1 vectors of phasor line voltages and currents have entries

$$\hat{\mathbf{V}}(z) = \begin{bmatrix} \hat{V}_G(z) \\ \hat{V}_R(z) \end{bmatrix} \tag{6.10a}$$

$$\hat{\mathbf{I}}(z) = \begin{bmatrix} \hat{I}_G(z) \\ \hat{I}_R(z) \end{bmatrix} \tag{6.10b}$$

and the 2×2 per-unit-length inductance and capacitance matrices have entries

$$\mathbf{L} = \begin{bmatrix} l_G & l_m \\ l_m & l_R \end{bmatrix} \quad \text{H/m} \tag{6.11a}$$

$$\mathbf{C} = \begin{bmatrix} c_G + c_m & -c_m \\ -c_m & c_R + c_m \end{bmatrix} \quad \text{F/m} \tag{6.11b}$$

The 2×2 per-unit-length resistance and conductance matrices are

$$\mathbf{R} = \begin{bmatrix} r_G + r_0 & r_0 \\ r_0 & r_R + r_0 \end{bmatrix} \quad \Omega/\text{m} \tag{6.12a}$$

$$\mathbf{G} = \begin{bmatrix} g_G + g_m & -g_m \\ -g_m & g_R + g_m \end{bmatrix} \quad \text{S/m} \tag{6.12b}$$

and the 2×2 per-unit-length impedance and admittance matrices are

$$\hat{\mathbf{Z}}(f) = \mathbf{R}(f) + j\omega \mathbf{L} \quad \Omega/\text{m} \tag{6.13a}$$

$$\hat{\mathbf{Y}}(f) = \mathbf{G}(f) + j\omega\mathbf{C} \qquad \text{S/m} \tag{6.13a}$$

The second-order uncoupled phasor MTL equations are obtained from (6.9) by differentiating one and substituting the other:

$$\boxed{\frac{d^2}{dz^2}\hat{\mathbf{V}}(z) = \hat{\mathbf{Z}}(f)\hat{\mathbf{Y}}(f)\hat{\mathbf{V}}(z)} \tag{6.14a}$$

$$\boxed{\frac{d^2}{dz^2}\hat{\mathbf{I}}(z) = \hat{\mathbf{Y}}(f)\hat{\mathbf{Z}}(f)\hat{\mathbf{I}}(z)} \tag{6.14b}$$

Note that in the MTL case, the order of multiplication of $\hat{\mathbf{Z}}$ and $\hat{\mathbf{Y}}$ matters. Our work in this chapter is focused on the solution of the second-order phasor MTL equations in (6.14).

6.2 CHARACTERIZATION OF CONDUCTOR AND DIELECTRIC LOSSES

As stated earlier, the line losses come about due to the resistance of the conductors and the conductance of the surrounding dielectric medium.

6.2.1 Conductor Losses and Skin Effect

At dc, the current is distributed uniformly over the cross sections of isolated conductors. As the frequency of the current in the conductors increases, the current tends to crowd to the surfaces of the conductors, thereby increasing their resistance, as shown in Fig. 6.3. The current tends to crowd into an annulus at the surface of the conductor of thickness a few skin depths. The skin depth is

$$\delta = \frac{1}{\sqrt{\pi f \mu \sigma}} = \frac{1}{\sqrt{\pi \mu_0 \sigma}}\frac{1}{\sqrt{f}}$$

Hence the high-frequency resistance increases as $\sqrt{f}$. For cylindrical circular conductors (commonly referred to as *wires*) that are isolated from other wires, the problem has been solved exactly, and their per-unit-length resistance is determined exactly. This can be approximated by two asymptotes, as shown in Fig. 6.4. However, as another wire is brought closer, the currents of the two wires tend to concentrate toward their facing sides; that is, the *proximity effect* becomes dominant, and the exact solution is invalidated.

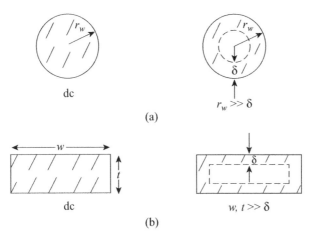

FIGURE 6.3. Skin effect in conductors.

In addition to their resistance, wires possess a per-unit-length *internal inductance* due to the magnetic field internal to the wire, which for an *isolated wire* carrying a dc current is

$$
\begin{aligned}
l_{i,\mathrm{dc}} &= \frac{\mu_0}{8\pi} \\
&= 0.5 \times 10^{-7} \\
&= 50\,\mathrm{nH/m} \\
&= 1.27\,\mathrm{nH/in}
\end{aligned}
\tag{6.15}
$$

This is in series with the *external inductances* that we have been considering for lossless lines that is due to the magnetic field external to the wire. The internal inductance of a wire is usually much less in magnitude than its

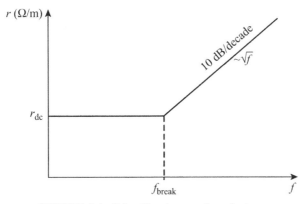

FIGURE 6.4. Skin effect losses of conductors.

external inductance and is therefore negligible. As the frequency of the current in the wire increases and the current tends to crowd to the wire surface, the per-unit-length resistance increases at a rate of $\sqrt{f}$, while this internal inductance *decreases* at a rate of $\sqrt{f}$. This gives further support to the notion that the per-unit-length internal inductance, l_i, can usually be neglected.

In the two frequency regions, the per-unit-length resistance and internal inductance for isolated wires having radii of r_w can be calculated approximately by

$$\left.\begin{aligned} r_{dc} &= \frac{1}{\sigma \pi r_w^2} \quad \Omega/m \\ l_{i,dc} &= \frac{\mu_0}{8\pi} \quad H/m \end{aligned}\right\} \quad r_w < 2\delta \quad (6.16a)$$

and

$$\left.\begin{aligned} r_{hf} &= \frac{1}{2\pi r_w \sigma \delta} = \frac{1}{2r_w}\sqrt{\frac{\mu_0}{\pi\sigma}}\sqrt{f} \quad \Omega/m \\ l_{i,hf} &= \frac{1}{4\pi r_w}\sqrt{\frac{\mu_0}{\pi\sigma}}\frac{1}{\sqrt{f}} \quad H/m \end{aligned}\right\} \quad r_w > 2\delta \quad (6.16b)$$

The high-frequency results in (6.16) derive from the fact that for wires the high-frequency resistance and the impedance of the high-frequency internal inductance approach an equality as $f \to \infty$: $r_{hf} = \omega l_{i,hf}$.

For conductors of rectangular cross section (commonly referred to as *lands*), found in and on PCBs, this is a very difficult problem and is solved numerically in an approximate fashion. Nevertheless, it can be approximated reasonably well by two asymptotes, as illustrated in Fig. 6.4. Again the two asymptotes join at a common break frequency f_0, where

$$\begin{aligned} \frac{wt}{w+t} &= k\delta \\ &= \frac{k}{\sqrt{\pi f_0 \mu \sigma}} \end{aligned} \quad (6.17)$$

The factor k is found numerically to be approximately $k \cong 1.3$. Hence the per-unit-length resistance of a PCB land is

$$\left.\begin{aligned} r_{dc} &= \frac{1}{\sigma wt} \quad \Omega/m \\ r_{hf} &= \frac{1}{k\sigma\delta(w+t)} \quad \Omega/m \end{aligned}\right\} \quad (6.18)$$

For wires it is true that the high-frequency resistance and the impedance of the high-frequency internal inductance approach an equality as $f \to \infty$: $r_{hf} = \omega l_{i,hf}$. For lands on a PCB, this is only approximately true. Hence we can obtain a simple and useful asymptotic representation for the per-unit-length conductor internal impedances for wires and PCB lands as

$$
\begin{aligned}
\hat{z}_i(\omega) &= r(\omega) + j\omega l_i(\omega) \\[2mm]
&= \begin{cases}
r_{dc}\left(1 + j\dfrac{f}{f_0}\right) & f < f_0 \\[4mm]
r_{dc}\sqrt{\dfrac{f}{f_0}}(1 + j) & f > f_0
\end{cases}
\end{aligned}
\tag{6.19}
$$

where f_0 is the break frequency where the dc and high-frequency asymptotes join. For wires we can obtain the break frequency from

$$
\begin{aligned}
r_w &= 2\delta \\[2mm]
&= \frac{2}{\sqrt{\pi f_0 \mu_0 \sigma}}
\end{aligned}
\tag{6.20}
$$

giving

$$
f_0 = \frac{4}{\pi r_w^2 \mu_0 \sigma}
\tag{6.21}
$$

We can show that at the break frequency f_0 the dc resistance and the dc internal inductive reactance are equal, $r_{dc} = \omega_0 l_{i,dc}$:

$$
\begin{aligned}
r_{dc} &= \frac{1}{\pi r_w^2 \sigma} \\[2mm]
&= 2\pi \underbrace{\frac{4}{\pi r_w^2 \mu_0 \sigma}}_{f_0} \times \underbrace{\frac{\mu_0}{8\pi}}_{l_{i,dc}} \\[2mm]
&= \omega_0 l_{i,dc}
\end{aligned}
\tag{6.22}
$$

and confirm that $r_{dc} = \omega_0 l_{i,dc}$ for wires. Similarly, we can confirm that because of the $\sqrt{f}$ behavior of r_{hf} and the $1/\sqrt{f}$ behavior of $l_{i,hf}$ and $r_{hf} = \omega l_{i,hf}$ and the low and high-frequency asymptotes join at f_0, the high-frequency behavior of conductors must satisfy

$$
r_{hf} = r_{dc}\sqrt{\frac{f}{f_0}}
\tag{6.23}
$$

and

$$\omega l_{i,\text{hf}} = r_{\text{dc}}\sqrt{\frac{f}{f_0}} \tag{6.24}$$

thereby confirming the representation in (6.19).

6.2.2 Dielectric Losses

As discussed in Chapter 3, dielectric losses result from the inability of the *bound-charge dipoles* in the dielectric to align instantaneously with the changing electric field. (See Fig. 3.18.) We model this phenomenon with a complex permittivity:

$$\boxed{\hat{\varepsilon} = \varepsilon' - j\varepsilon''} \tag{6.25}$$

and the real part, $\varepsilon' = \varepsilon_r \varepsilon_0$, is the usual permittivity of the dielectric. The imaginary part, ε'', is associated with the losses of the dielectric. The dielectric losses are given at various frequencies in handbooks in terms of a *loss tangent* as the ratio of the imaginary and real parts of this complex permittivity:

$$\boxed{\tan\theta = \frac{\varepsilon''}{\varepsilon'}} \tag{6.26}$$

and the complex permittivity can be written as

$$\boxed{\hat{\varepsilon} = \varepsilon'(1 - j\tan\theta)} \tag{6.27}$$

Substituting (6.25) into Ampère's law gives

$$\begin{aligned}
\nabla \times \vec{H} &= \sigma\vec{E} + j\omega\hat{\varepsilon}\vec{E} \\
&= (\sigma + \omega\varepsilon'')\vec{E} + j\omega\varepsilon'\vec{E}
\end{aligned} \tag{6.28}$$

Hence we obtain an *effective conductivity* of the dielectric (which obviously represents dielectric loss) as

$$\sigma_{\text{eff}} = \sigma + \omega\varepsilon'' \tag{6.29}$$

For typical dielectrics, the Ohmic conductivity σ is so small that it can be neglected, and the effect of the bound charges is the dominant effect. Hence the effective conductivity of most dielectrics is just

$$\boxed{\begin{aligned} \sigma_{\text{eff}} &= \omega\varepsilon'' \\ &= \omega\varepsilon'\tan\theta \end{aligned}}$$

(6.30)

Homogeneous Medium   We first consider a *homogeneous medium* sur-rounding the conductors. For a homogeneous medium surrounding the conductors, the per-unit-length capacitance matrix, **C**, depends only on the line cross-sectional dimensions and the parameters of the surrounding homogeneous medium as $\mathbf{C} = \varepsilon'\mathbf{K}$, where **K** is a constant matrix and $\varepsilon' = \varepsilon_r\varepsilon_0$. Similarly, for a homogeneous medium we expect the per-unit-length conductance matrix to depend on the same **K**, $\mathbf{G} = \sigma_{\text{eff}}\mathbf{K}$, where the 2×2 constant matrix **K** is the same in both cases. Hence for a *homogeneous medium*, the per-unit-length conductance matrix, **G**, and the per-unit-length capacitance matrix, **C**, are related as

$$\mathbf{G} = \frac{\sigma_{\text{eff}}}{\varepsilon'}\mathbf{C}$$

(6.31)

For a *homogeneous medium* we can determine the per-unit-length conductance matrix in terms of the per-unit-length capacitance matrix and the loss tangent of the material by substituting (6.30) as

$$\boxed{\mathbf{G} = \omega\tan\theta\,\mathbf{C}}$$

(6.32)

Although this seems to imply that the entries in the per-unit-length conductance **G** increase *linearly* with frequency, this is not the case. The loss tangent has numerous frequency bands of *resonance*, much like a bandpass filter. A common way of characterizing this frequency dependence is with the *Debye model*. This represents the complex permittivity with a series as

$$\begin{aligned} \hat{\varepsilon}(\omega) &= \varepsilon'(\omega) - j\varepsilon''(\omega) \\ &= \varepsilon'(\omega)(1 - j\tan\theta) \\ &= \varepsilon_{\text{hf}} + \sum_{i=1}^{N} \frac{K_i}{1 + j\omega\tau_i} \end{aligned}$$

(6.33)

where the N constants K_i and time constants τ_i are to be found for the particular dielectric experimentally. At dc, $\varepsilon_r\varepsilon_0 = \varepsilon_{\text{hf}} + \sum_{i=1}^{N} K_i$. Noting that

$$\frac{K_i}{1 + j\omega\tau_i} \times \frac{1 - j\omega\tau_i}{1 - j\omega\tau_i} = \frac{K_i - jK_i\omega\tau_i}{1 + \omega^2\tau_i^2}$$

we obtain the real and imaginary parts of the complex permittivity $\hat{\varepsilon}(\omega)$ as

$$\varepsilon'(\omega) = \varepsilon_{hf} + \sum_{i=1}^{N} \frac{K_i}{1 + \omega^2 \tau_i^2} \qquad (6.34a)$$

$$\varepsilon''(\omega) = \sum_{i=1}^{N} \frac{K_i \omega \tau_i}{1 + \omega^2 \tau_i^2} \qquad (6.34b)$$

The imaginary part in (6.34) for ε'' (representing the dielectric loss) has the same form as a bandpass filter and represents the various lossy resonant regions. The real part, ε', in (6.34a) decreases from $\varepsilon' = \varepsilon_r \varepsilon_0$ as $f \to 0$ to $\varepsilon' \to \varepsilon_0$ as $f \to \infty$, much like a low-pass filter. Hence the capacitance matrix is also frequency dependent, $\mathbf{C}(\omega)$.

Inhomogeneous Medium For a *homogeneous medium* surrounding the line conductors, such as a coupled stripline, the development above is straightforward since there is only one permittivity and one loss tangent for the homogeneous region. But for an *inhomogeneous medium* surrounding the conductors, such as in the case of (a) insulations on wires or (b) a coupled microstripline or a common PCB, this is not so straightforward. For an inhomogeneous medium the determination of $\mathbf{G}$ is more difficult. Once again we represent *each region* with its complex permittivity in terms of its permittivity and its loss tangent as

$$\hat{\varepsilon}_i = \varepsilon_i'(1 - j \tan \theta_i) \qquad (6.35)$$

or with its Debye model as in (6.34). Again we employ a numerical approximation program which determines $\mathbf{C}$, such as RIBBON.FOR, PCB.FOR, MSTRP.FOR, or STRPLINE.FOR, and representing each region with its complex permittivity as in (6.35) or with the Debye model as in (6.34), we will obtain a *complex* per-unit-length capacitance matrix [which is also frequency dependent because $\hat{\varepsilon}_i(\omega)$ is also frequency dependent]:

$$\hat{\mathbf{C}}(\omega) = \mathbf{C}_R(\omega) + j\mathbf{C}_I(\omega) \qquad (6.36)$$

The per-unit-length admittance matrix is

$$\hat{\mathbf{Y}}(\omega) = \mathbf{G}(\omega) + j\omega\mathbf{C}(\omega) \qquad (6.37)$$

Substitute (6.36) into

$$\hat{\mathbf{Y}}(\omega) = j\omega\hat{\mathbf{C}}(\omega)$$
$$= \underbrace{-\omega\mathbf{C}_I}_{\mathbf{G}} + j\omega\underbrace{\mathbf{C}_R}_{\mathbf{C}} \qquad (6.38)$$

Hence the imaginary part of $\hat{\mathbf{C}}$ in (6.36) is related to the desired per-unit-length conductance matrix, $\mathbf{G}$, and the real part of (6.36) is related to the desired per-unit-length capacitance matrix, $\mathbf{C}$, as

$$\boxed{\mathbf{C}(\omega) = \mathbf{C}_R(\omega)} \qquad (6.39a)$$

$$\boxed{\mathbf{G}(\omega) = -\omega\mathbf{C}_I(\omega)} \qquad (6.39b)$$

So determining $\mathbf{G}$ and $\mathbf{C}$ is very difficult and must be done numerically for an *inhomogeneous medium* (and at every desired frequency).

6.3 SOLUTION OF THE PHASOR (FREQUENCY-DOMAIN) TRANSMISSION-LINE EQUATIONS FOR A THREE-CONDUCTOR LOSSY LINE

We will (arbitrarily) choose to solve the second-order MTL equations for the phasor current in (6.14):

$$\frac{d^2}{dz^2}\hat{\mathbf{I}}(z) = \hat{\mathbf{Y}}(f)\hat{\mathbf{Z}}(f)\hat{\mathbf{I}}(z) \qquad (6.14b)$$

Again we will transform the actual line currents to the *mode currents* with a *similarity transformation* as

$$\hat{\mathbf{I}}(z) = \hat{\mathbf{T}}\hat{\mathbf{I}}_{\text{mode}}(z) \qquad (6.40)$$

where the 2×2 similarity transformation $\hat{\mathbf{T}}$ (which is, in general, frequency dependent and complex valued) is to be determined. Substituting (6.40) into (6.14b) gives

$$\frac{d^2}{dz^2}\hat{\mathbf{I}}_{\text{mode}}(z) = \hat{\mathbf{T}}^{-1}\hat{\mathbf{Y}}\hat{\mathbf{Z}}\hat{\mathbf{T}}\hat{\mathbf{I}}_{\text{mode}}(z) \qquad (6.41)$$

Define the 2×2 *propagation matrix* as

$$\hat{\gamma}^2 = \hat{\mathbf{T}}^{-1}\hat{\mathbf{Y}}\hat{\mathbf{Z}}\hat{\mathbf{T}} \qquad (6.42)$$

If we can find a $\hat{\mathbf{T}}$ that will diagonalize the product $\hat{\mathbf{Y}}\hat{\mathbf{Z}}$ (at each desired frequency) where $\hat{\gamma}^2$ is a diagonal matrix,

$$\hat{\gamma}^2 = \begin{bmatrix} \hat{\gamma}_G^2 & 0 \\ 0 & \hat{\gamma}_R^2 \end{bmatrix} \tag{6.43}$$

the mode equations in (6.41) will be *uncoupled*, whose general solution is simple and was obtained in Chapter 3 as

$$\hat{\mathbf{I}}_{\text{mode}} = \mathbf{e}^{-\hat{\gamma}z}\hat{\mathbf{I}}_m^+ - \mathbf{e}^{\hat{\gamma}z}\hat{\mathbf{I}}_m^- \tag{6.44}$$

The 2×1 vectors $\hat{\mathbf{I}}_m^+$ and $\hat{\mathbf{I}}_m^-$ have entries yet to be determined by applying the terminal constraints at the source, $z = 0$, and at the load, $z = \mathscr{L}$, as

$$\hat{\mathbf{I}}_m^+ = \begin{bmatrix} I_{mG}^+ \\ I_{mR}^+ \end{bmatrix} \tag{6.45a}$$

$$\hat{\mathbf{I}}_m^- = \begin{bmatrix} I_{mG}^- \\ I_{mR}^- \end{bmatrix} \tag{6.45b}$$

The 2×2 exponential matrices are diagonal:

$$\mathbf{e}^{-\hat{\gamma}z} = \begin{bmatrix} e^{-\hat{\gamma}_G z} & 0 \\ 0 & e^{-\hat{\gamma}_R z} \end{bmatrix} \tag{6.46a}$$

$$\mathbf{e}^{\hat{\gamma}z} = \begin{bmatrix} e^{\hat{\gamma}_G z} & 0 \\ 0 & e^{\hat{\gamma}_R z} \end{bmatrix} \tag{6.46b}$$

The 2×2 matrix $\hat{\gamma}^2$ is diagonal and has the (two) *eigenvalues* $\hat{\gamma}_G^2$ and $\hat{\gamma}_R^2$ of the product $\hat{\mathbf{Y}}\hat{\mathbf{Z}}$ on the main diagonal:

$$\begin{aligned} \hat{\gamma}^2 &= \hat{\mathbf{T}}^{-1}\hat{\mathbf{Y}}\hat{\mathbf{Z}}\hat{\mathbf{T}} \\ &= \begin{bmatrix} \hat{\gamma}_G^2 & 0 \\ 0 & \hat{\gamma}_R^2 \end{bmatrix} \end{aligned} \tag{6.47}$$

This is a standard *eigenvalue–eigenvector* problem in linear algebra. We use a subroutine EIGCC from the International Mathematical and Statistical Library (IMSL) that determines the eigenvalues $\hat{\gamma}_G^2$ and $\hat{\gamma}_R^2$ of $\hat{\mathbf{Y}}\hat{\mathbf{Z}}$ and eigenvectors of $\hat{\mathbf{Y}}\hat{\mathbf{Z}}$ (which are the columns of $\hat{\mathbf{T}}$) of any square complex-valued matrix. Once the undetermined vectors $\hat{\mathbf{I}}_m^+$ and $\hat{\mathbf{I}}_m^-$ are determined by applying the terminal constraints at the source, $z = 0$, and at the load, $z = \mathscr{L}$, the actual line currents are determined from (6.40):

$$\hat{\mathbf{I}}(z) = \hat{\mathbf{T}}\hat{\mathbf{I}}_{\text{mode}}(z)$$
$$= \hat{\mathbf{T}}\left(e^{-\hat{\gamma}z}\hat{\mathbf{I}}_m^+ - e^{\hat{\gamma}z}\hat{\mathbf{I}}_m^-\right) \tag{6.48a}$$

and the actual line voltages are determined from (6.9):

$$\hat{\mathbf{V}}(z) = -\hat{\mathbf{Y}}^{-1}\frac{d}{dz}\hat{\mathbf{I}}(z)$$
$$= \underbrace{\hat{\mathbf{Y}}^{-1}\hat{\mathbf{T}}\hat{\gamma}\hat{\mathbf{T}}^{-1}}_{\hat{\mathbf{Z}}_C}\hat{\mathbf{T}}\left(e^{-\hat{\gamma}z}\hat{\mathbf{I}}_m^+ + e^{\hat{\gamma}z}\hat{\mathbf{I}}_m^-\right) \tag{6.48b}$$

where the 2×2 *characteristic impedance matrix* is defined as

$$\hat{\mathbf{Z}}_C = \hat{\mathbf{Y}}^{-1}\hat{\mathbf{T}}\hat{\gamma}\hat{\mathbf{T}}^{-1} \tag{6.48c}$$

The 2×1 vectors of undetermined constants, $\hat{\mathbf{I}}_m^+$ and $\hat{\mathbf{I}}_m^-$, are determined by applying the terminal constraints at the source, $z = 0$, and at the load, $z = \mathscr{L}$. How shall we characterize these terminations? The obvious way is with a *generalized Thévenin equivalent*, which involves matrices as

$$\hat{\mathbf{V}}(0) = \hat{\mathbf{V}}_S - \hat{\mathbf{Z}}_S\hat{\mathbf{I}}(0) \tag{6.49a}$$

$$\hat{\mathbf{V}}(\mathscr{L}) = \hat{\mathbf{V}}_L + \hat{\mathbf{Z}}_L\hat{\mathbf{I}}(\mathscr{L}) \tag{6.49b}$$

where

$$\hat{\mathbf{V}}(z) = \begin{bmatrix} \hat{V}_G(z) \\ \hat{V}_R(z) \end{bmatrix} \tag{6.50a}$$

and

$$\hat{\mathbf{I}}(z) = \begin{bmatrix} \hat{I}_G(z) \\ \hat{I}_R(z) \end{bmatrix} \tag{6.50b}$$

For example, for the general three-conductor representation in Fig. 4.1,

$$\hat{\mathbf{V}}_S = \begin{bmatrix} \hat{V}_S \\ 0 \end{bmatrix} \tag{6.51a}$$

$$\hat{\mathbf{Z}}_S = \begin{bmatrix} R_S & 0 \\ 0 & R_{NE} \end{bmatrix} \tag{6.51b}$$

$$\hat{\mathbf{V}}_L = \begin{bmatrix} 0 \\ 0 \end{bmatrix} \tag{6.51c}$$

$$\hat{\mathbf{Z}}_L = \begin{bmatrix} R_L & 0 \\ 0 & R_{FE} \end{bmatrix} \tag{6.51d}$$

Substituting (6.49) into (6.48) gives four equations in four unknowns (the unknowns in $\hat{\mathbf{I}}_m^+$ and $\hat{\mathbf{I}}_m^-$) to be determined:

$$\begin{bmatrix} (\hat{\mathbf{Z}}_C + \hat{\mathbf{Z}}_S)\hat{\mathbf{T}} & (\hat{\mathbf{Z}}_C - \hat{\mathbf{Z}}_S)\hat{\mathbf{T}} \\ (\hat{\mathbf{Z}}_C - \hat{\mathbf{Z}}_L)\hat{\mathbf{T}}e^{-\hat{\gamma}\mathscr{L}} & (\hat{\mathbf{Z}}_C + \hat{\mathbf{Z}}_L)\hat{\mathbf{T}}e^{\hat{\gamma}\mathscr{L}} \end{bmatrix} \begin{bmatrix} \hat{\mathbf{I}}_m^+ \\ \hat{\mathbf{I}}_m^- \end{bmatrix} = \begin{bmatrix} \hat{\mathbf{V}}_S \\ \hat{\mathbf{V}}_L \end{bmatrix} \tag{6.52}$$

We use a subroutine LEQT1C from the International Mathematical and Statistical Library (IMSL) to solve the complex simultaneous equations in (6.52). Once the unknowns in $\hat{\mathbf{I}}_m^+$ and $\hat{\mathbf{I}}_m^-$, which are the simultaneous solution of (6.52), are determined, the line voltages and currents can be determined at any point on the line from (6.48) and specifically at the source, $z = 0$, and at the load, $z = \mathscr{L}$. The crosstalk voltages are then determined from (6.48) as $\hat{V}_{NE} = \hat{V}_R(0)$ and $\hat{V}_{FE} = \hat{V}_R(\mathscr{L})$.

EXAMPLE

As an example, consider the three-conductor line shown in Fig. 6.5. This was also considered in Fig. 5.4, but here the 0 to 100% rise and fall times are reduced from 6.25 ns to 1.25 ns.

First we compute the frequency-domain transfer functions. For $\hat{V}_S = 1\angle 0°$ this gives the near- and far-end phasor transfer functions (magnitude and angle) $\hat{V}_{NE}/\hat{V}_S$ and $\hat{V}_{FE}/\hat{V}_S$.

The MTL Computer Program (MTL.FOR, MTL.EXE)   MTL.FOR is a FORTRAN program that first diagonalizes $\hat{\mathbf{Y}}\hat{\mathbf{Z}}$ as in (6.42) and then determines the frequency-domain solution of equations (6.52). The

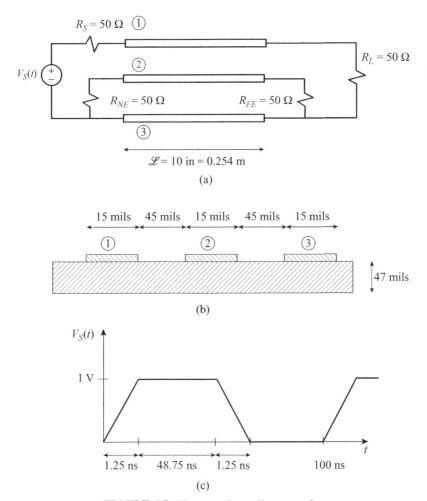

FIGURE 6.5. Three-conductor line example.

program MTL.EXE is the compiled and executable file. The input files to MTL.EXE are MTL.IN and PUL.DAT, and the frequencies to be analyzed are in FREQ.IN. MTL.EXE reads the input data stored in the files MTL.IN, PUL.DAT, and FREQ.IN and produces as output the file MTL.OUT, which contains the phasor (magnitude and phase) of the near- and far-end crosstalk voltages, $\hat{V}_{NE} = V_{NE} \angle \theta_{NE}$ and $\hat{V}_{FE} = V_{FE} \angle \theta_{FE}$. Although our concentration will be on three-conductor lines, the program handles $n + 1$ conductors, where $n + 1$ is any number up to the array dimensions of MTL.FOR. MTL. FOR includes the conductor losses but neglects dielectric losses.

For the problem in Fig. 6.5, the input file MTL.IN is

```
3                       =Total Number of Conductors (N+1)
0.254                   =Total Line Length (m)
2                       =Output Conductor for Terminal
                         Voltages
1.2910          =Reference Conductor dc Resistance
                 (ohms/m)
6.977E6        =Reference Conductor Resistance Break
                Frequency(Hz)
1    1.2910     =I,dc Conductor(I) Resistance (ohms/m)
1    6.977E6    =I,Conductor(I) Resistance Break
                 Frequency(Hz)
2    1.2910     =I,dc Conductor(I) Resistance (ohms/m)
2    6.977E6    =I,Conductor(I) Resistance Break
                 Frequency (Hz)
1       1.0    0.000    =I,VSREAL(I),VSIMAG(I)
2       0.0    0.000    =I,VSREAL(I),VSIMAG(I)
1    1  50.0   0.000    =I,J,ZSREAL(I,J),ZSIMAG(I,J)
1    2  0.0    0.000    =I,J,ZSREAL(I,J),ZSIMAG(I,J)
2    2  50.0   0.000    =I,J,ZSREAL(I,J),ZSIMAG(I,J)
1       0.0    0.000    =I,VLREAL(I),VLIMAG(I)
2       0.0    0.000    =I,VLREAL(I),VLIMAG(I)
1    1  50.0   0.000    =I,J,ZLREAL(I,J),ZLIMAG(I,J)
1    2  0.0    0.000    =I,J,ZLREAL(I,J),ZLIMAG(I,J)
2    2  50.0   0.000    =I,J,ZLREAL(I,J),ZLIMAG(I,J)
```

The dc per-unit-length resistances of the lands (1.38 mils × 15 mils) are

$$r_{dc} = \frac{1}{\sigma w t}$$
$$= 1.291 \ \Omega/m$$

and the break frequency between the dc and the high-frequency asymptotes in Fig. 6.4 is 6.977 MHz, where

$$\frac{wt}{w+t} = 1.3\delta$$
$$= \frac{1.3}{\sqrt{\pi f_0 \mu_0 \sigma}}$$

Note that the generalized Thévenin termination descriptions in (6.51) are described at the end of this file.

MTL.FOR also reads the L and C matrices from PUL.DAT:

```
1    1      1.38314E-06        =L(   1,   1)
1    2      6.91570E-07        =L(   1,   2)
2    2      1.10706E-06        =L(   2,   2)
1    1      2.96949E-11        =C(   1,   1)
1    2     -2.02619E-11        =C(   1,   2)
2    2      4.05238E-11        =C(   2,   2)
1    1      1.16983E-11        =C0(   1,   1)
1    2     -7.30777E-12        =C0(   1,   2)
2    2      1.46155E-11        =C0(   2,   2)


NUMBER OF LANDS=    3
NUMBER OF DIVISIONS PER LAND=    30
REFERENCE LAND=    3
LAND WIDTH (mils)=    1.500E+01
EDGE-TO-EDGE SEPARATION (mils)=    4.500E+01
BOARD THICKNESS (mils)=    4.700E+01
RELATIVE DIELECTRIC CONSTANT=    4.700E+00
```

The frequencies for the analysis are read from FREQ.IN. To compute and give enough detail to the transfer functions from 10 kHz to 1 GHz requires a large number of frequencies spaced, for example, 10 kHz, in FREQ.IN:

```
1.000E+04
2.000E+04
3.000E+04
4.000E+04
5.000E+04
6.000E+04
7.000E+04
8.000E+04
9.000E+04
1.000E+05
1.100E+05
1.200E+05
   ⋮
1.000E+09
```

Figures 6.6 and 6.7 show the magnitude and phase computed for the transfer functions from 10 kHz to 1 GHz for both the lossless case computed with

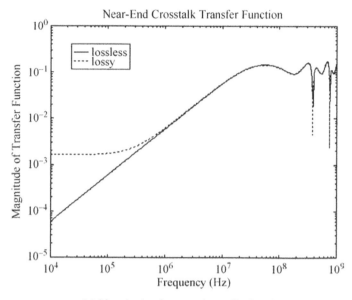

(a) Magnitude of near-end transfer function.

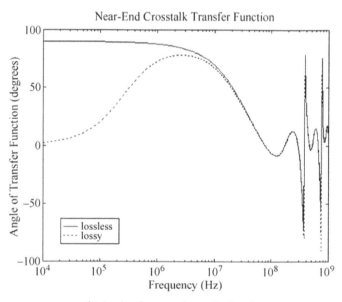

(b) Angle of near-end transfer function.

FIGURE 6.6. The (a) magnitude and (b) angle of the near-end frequency-domain transfer function $\hat{V}_{NE}/\hat{V}_S$ for the circuit of Fig. 6.5 computed for lossless lines with the PSPICE subcircuit model generated by SPICEMTL.FOR, and computed including conductor losses with MTL.FOR.

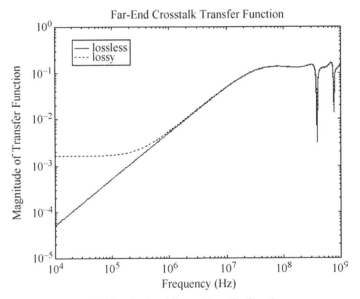

(a) Magnitude of far-end transfer function.

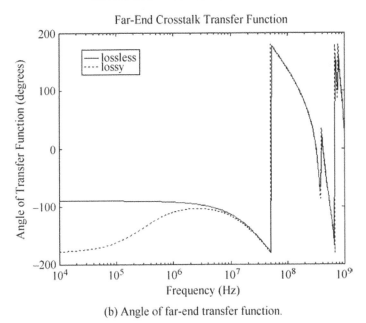

(b) Angle of far-end transfer function.

FIGURE 6.7. The (a) magnitude and (b) angle of the far end frequency-domain transfer function $\hat{V}_{FE}/\hat{V}_S$ for the circuit of Fig. 6.5 computed for lossless lines with the PSPICE subcircuit model generated by SPICEMTL.FOR, and computed including conductor losses with MTL.FOR.

PSPICE using the subcircuit model for the lossless line obtained with SPICEMTL.FOR and including the conductor losses computed by solving (6.52) using MTL.FOR. The output file for MTL.FOR is MTL.OUT and contains the near- and far-end crosstalk phasor voltages (magnitude and phase) at each frequency specified for the analysis in FREQ.IN. For $\hat{V}_S = 1 \angle 0°$ this gives the near- and far-end phasor transfer functions (magnitude and angle) $\hat{V}_{NE}/\hat{V}_S$ and $\hat{V}_{FE}/\hat{V}_S$ at each of the frequencies specified in FREQ.IN, which are plotted in Figs. 6.6 and 6.7.

6.4 COMMON-IMPEDANCE COUPLING

There is a very simple but very important method of crosstalk that relies simply on Ohm's law in addition to the electric and magnetic field coupling between the two circuits. The generator and receptor circuits share a common conductor, which we have designated as the zeroth or reference conductor for the line voltages. This is illustrated in Fig. 6.8.

The generator line current is approximately $I_G \cong V_S(t)/(R_S + R_L)$, and virtually all of this goes through the reference conductor generating a voltage between its two endpoints of

$$V_{CI} \cong r_0 \mathscr{L} I_G$$

$$\cong r_0 \mathscr{L} \frac{V_S(t)}{R_S + R_L}$$

This voltage is voltage-divided among the near- and far-end terminations to give common impedance components of the crosstalk voltages of

$$\boxed{V_{NE,CI} = \frac{R_{NE}}{R_{NE} + R_{FE}} r_0 \mathscr{L} \frac{V_S(t)}{R_S + R_L}}$$ (6.53a)

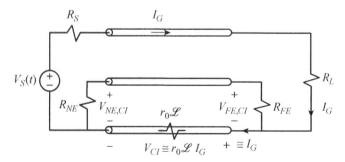

FIGURE 6.8. Common-impedance coupling.

and

$$V_{FE,CI} = -\frac{R_{FE}}{R_{NE} + R_{FE}} r_0 \mathscr{L} \frac{V_S(t)}{R_S + R_L} \qquad (6.53b)$$

For the problem in Fig. 6.5, these are $V_{NE,CI} = 1.64\,\text{mV}$ and $V_{FE,CI} = -1.64\,\text{mV}$. We have seen in previous chapters that for a *lossless line*, as $f \to 0$, the crosstalk voltages also go to zero *linearly*. These common-impedance-generated voltages place a "floor" on how small the magnitudes of the phasor crosstalk voltages can become as $f \to 0$. Observe that these common-impedance floor voltages are easily calculated. The *magnitudes* of the frequency-response transfer functions in Fig. 6.6(a) and Fig. 6.7(a) clearly show a floor of $\hat{V}_{NE,CI}/\hat{V}_S = 1.64 \times 10^{-3}$ and $\hat{V}_{FE,CI}/\hat{V}_S = -1.64 \times 10^{-3}$.

6.5 THE TIME-DOMAIN TO FREQUENCY-DOMAIN METHOD

In Chapter 1 we showed a simple method for incorporating these frequency-dependent line parameters into the time-domain results. This was simply the superposition of the time-domain responses of the frequency-domain (phasor) responses to the Fourier components of the input source waveform, $V_S(t)$, as illustrated in Fig. 6.9.

The problem must be represented as a single-input, $V_S(t)$, single-output, $V_{NE}(t)$ or $V_{FE}(t)$, linear system and is obtained by embedding the terminations into the system as illustrated in Fig. 6.10(a). The frequency-domain responses to the Fourier harmonics of the source waveform, $V_S(t)$, is illustrated in Fig. 6.10(b). This simple process is referred to as the *time-domain to*

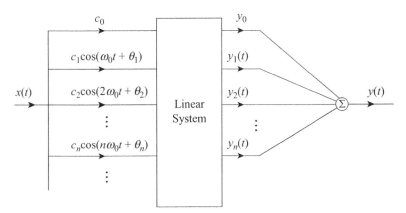

FIGURE 6.9. Using superposition to determine the time-domain response of a linear system.

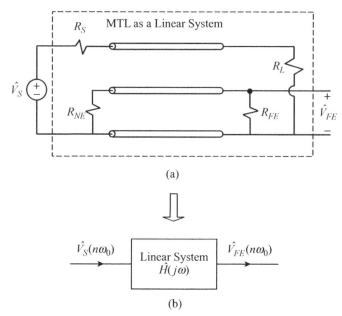

(a)

⇩

(b)

FIGURE 6.10. MTL as a linear system.

frequency-domain (TDFD) *method* or "a poor man's FFT":

$$
V_{NE,FE}(t) = c_0 H_{NE,FE}(0) + \sum_{n=1}^{N} |c_n| |\hat{H}_{NE,FE}(j\omega)| \\
\times \cos\left(n\omega_0 t + \angle c_n + \angle \hat{H}_{NE,FE}(j\omega)\right)
\tag{6.54}
$$

Since superposition is used, the system (and terminations) must be a linear one (and in steady state).

Again MTL.FOR can be used to compute the frequency-domain responses at each of the harmonic Fourier components of $V_S(t)$, but the frequencies of the analysis in FREQ.IN must be those of the Fourier components of $V_S(t)$. For this problem $f_0 = 10$ MHz and we will superimpose 50 harmonics. So the transfer function must be be computed at 50 harmonics and FREQ.IN must be changed from the previous transfer function problem.

The frequencies of the analysis are input to MTL.FOR in FREQ.IN. In preparation for the superposition in TIMEFREQ.FOR of the responses to fifty 10-MHz harmonics, the computation frequencies in FREQ.IN are changed to

```
1.0E7
2.0E7
3.0E7
```

```
4.0E7
5.0E7
6.0E7
7.0E7
8.0E7
9.0E7
10.0E7
11.0E7
12.0E7
13.0E7
14.0E7
15.0E7
16.0E7
17.0E7
18.0E7
19.0E7
20.0E7
21.0E7
22.0E7
23.0E7
24.0E7
25.0E7
26.0E7
27.0E7
28.0E7
29.0E7
30.0E7
31.0E7
32.0E7
33.0E7
34.0E7
35.0E7
36.0E7
37.0E7
38.0E7
39.0E7
40.0E7
41.0E7
42.0E7
43.0E7
44.0E7
45.0E7
46.0E7
```

```
47.0E7
48.0E7
49.0E7
50.0E7
```

The output file for MTL.FOR is MTL.OUT and contains the near- and far-end crosstalk phasor voltages (magnitude and phase) at each frequency specified for the analysis in FREQ.IN. For $\hat{V}_S = 1 \angle 0°$ **this gives the near- and far-end phasor transfer functions (magnitude and angle)** $\hat{V}_{NE}/\hat{V}_S$ **and** $\hat{V}_{FE}/\hat{V}_S$. For the problem in Fig. 6.5, the output file is MTL.OUT:

```
TOTAL NUMBER OF CONDUCTORS= 3
LINE LENGTH (METERS)= 2.540E-01
OUTPUT CONDUCTOR= 2

DC  RESISTANCE  OF  REFERENCE  CONDUCTOR  (OHMS/M)=
1.291E+00
BREAK FREQUENCY (HZ)= 6.977E+06

I        RDC            BREAK FREQUENCY
1      1.291E+00          6.977E+06
2      1.291E+00          6.977E+06

TERMINAL NETWORK CHARACTERIZATION
I J  VSR    VSI     ZSR     ZSI     VLR     VLI     ZLR     ZLI
1 1 1.0E+00 0.0E+00 5.0E+01 0.0E+00 0.0E+00 0.0E+00 5.0E+01 0.0E+00
2 2 0.0E+00 0.0E+00 5.0E+01 0.0E+00 0.0E+00 0.0E+00 5.0E+01 0.0E+00
1 2         0.0E+00 0.0E+00         0.0E+00 0.0E+00

PER-UNIT-LENGTH PARAMETERS
L(  1,  1)=   1.38314E-06
L(  1,  2)=   6.91570E-07
L(  2,  2)=   1.10706E-06
C(  1,  1)=   2.96949E-11
C(  1,  2)=  -2.02619E-11
C(  2,  2)=   4.05238E-11

OUTPUT CONDUCTOR= 2

FREQUENCY     VNEMAG       VNEANG       VFEMAG      VFEANG
1.000E+07  5.718E-02   6.678E+01  4.975E-02  -1.167E+02
```

2.000E+07	9.873E-02	4.936E+01	8.643E-02	−1.372E+02
3.000E+07	1.239E-01	3.534E+01	1.098E-01	−1.544E+02
4.000E+07	1.370E-01	2.436E+01	1.235E-01	−1.687E+02
5.000E+07	1.421E-01	1.572E+01	1.311E-01	1.792E+02
6.000E+07	1.424E-01	8.875E+00	1.349E-01	1.687E+02
7.000E+07	1.396E-01	3.439E+00	1.366E-01	1.594E+02
8.000E+07	1.350E-01	−8.243E-01	1.369E-01	1.509E+02
9.000E+07	1.292E-01	−4.062E+00	1.365E-01	1.432E+02
1.000E+08	1.230E-01	−6.362E+00	1.356E-01	1.361E+02
1.100E+08	1.166E-01	−7.774E+00	1.345E-01	1.293E+02
1.200E+08	1.104E-01	−8.327E+00	1.335E-01	1.230E+02
1.300E+08	1.047E-01	−8.046E+00	1.325E-01	1.169E+02
1.400E+08	9.980E-02	−6.970E+00	1.317E-01	1.110E+02
1.500E+08	9.586E-02	−5.174E+00	1.311E-01	1.054E+02
1.600E+08	9.308E-02	−2.782E+00	1.307E-01	9.980E+01
1.700E+08	9.163E-02	2.130E-02	1.307E-01	9.432E+01
1.800E+08	9.158E-02	2.998E+00	1.309E-01	8.889E+01
1.900E+08	9.295E-02	5.891E+00	1.315E-01	8.346E+01
2.000E+08	9.569E-02	8.457E+00	1.323E-01	7.799E+01
2.100E+08	9.968E-02	1.050E+01	1.335E-01	7.245E+01
2.200E+08	1.047E-01	1.189E+01	1.350E-01	6.680E+01
2.300E+08	1.107E-01	1.256E+01	1.367E-01	6.099E+01
2.400E+08	1.174E-01	1.246E+01	1.387E-01	5.498E+01
2.500E+08	1.245E-01	1.160E+01	1.408E-01	4.872E+01
2.600E+08	1.318E-01	9.967E+00	1.431E-01	4.215E+01
2.700E+08	1.390E-01	7.565E+00	1.454E-01	3.522E+01
2.800E+08	1.459E-01	4.376E+00	1.476E-01	2.785E+01
2.900E+08	1.519E-01	3.596E-01	1.494E-01	1.994E+01
3.000E+08	1.566E-01	−4.556E+00	1.505E-01	1.137E+01
3.100E+08	1.592E-01	−1.049E+01	1.505E-01	1.992E+00
3.200E+08	1.586E-01	−1.763E+01	1.485E-01	−8.424E+00
3.300E+08	1.533E-01	−2.627E+01	1.435E-01	−2.019E+01
3.400E+08	1.409E-01	−3.683E+01	1.335E-01	−3.372E+01
3.500E+08	1.181E-01	−4.977E+01	1.155E-01	−4.948E+01
3.600E+08	8.137E-02	−6.514E+01	8.609E-02	−6.741E+01
3.700E+08	3.132E-02	−7.909E+01	4.464E-02	−8.433E+01
3.800E+08	2.395E-02	6.203E+01	8.894E-03	−3.569E+00
3.900E+08	6.910E-02	5.108E+01	4.644E-02	3.114E+01
4.000E+08	9.935E-02	3.742E+01	7.685E-02	1.708E+01
4.100E+08	1.164E-01	2.623E+01	9.683E-02	3.302E+00
4.200E+08	1.243E-01	1.747E+01	1.093E-01	−8.745E+00
4.300E+08	1.264E-01	1.068E+01	1.169E-01	−1.925E+01

```
4.400E+08  1.250E-01   5.488E+00  1.215E-01  −2.855E+01
4.500E+08  1.216E-01   1.625E+00  1.241E-01  −3.692E+01
4.600E+08  1.170E-01  −1.081E+00  1.257E-01  −4.456E+01
4.700E+08  1.120E-01  −2.741E+00  1.265E-01  −5.165E+01
4.800E+08  1.069E-01  −3.433E+00  1.269E-01  −5.828E+01
4.900E+08  1.022E-01  −3.221E+00  1.272E-01  −6.457E+01
5.000E+08  9.826E-02  −2.184E+00  1.275E-01  −7.059E+01
```

***The TIMEFREQ Computer Program (TIMEFREQ.FOR, TIMEFREQ.
EXE)*** TIMEFREQ.FOR is a FORTRAN program that performs the
superposition of the responses to the harmonics of $V_S(t)$ as in (6.54). The
program TIMEFREQ.EXE is the compiled and executable file. The input files
to TIMEFREQ.EXE are TIMEFREQ.IN, and the transfer functions at each
harmonic are in MTLFREQ.DAT.

The header material in MTL.OUT is stripped off and the output renamed
MTLFREQ.DAT in preparation for its submittal (as the phasor transfer
function) to the compiled and executable file TIMEFREQ.EXE. Other than
the input file MTLFREQ.DAT, which contains the frequency-domain transfer
functions, the other input file to TIMEFREQ.EXE is TIMEFREQ.IN:

```
50                 =NUMBER OF FREQUENCIES (HARMONICS)
1.0                =PEAK PULSE LEVEL
1.E7               =REPETITION FREQUENCY OF PULSE TRAIN
0.5                =DUTY CYCLE OF PULSE TRAIN
1.25E-9             =PULSE RISE/FALL TIME
100.E-9             =FINAL SOLUTION TIME
0.1E-9              =SOLUTION TIME INCREMENT
1.618E-3            =DC VALUE OF TRANSFER FUNCTION, H(0)
```

TIMEFREQ.IN requires as input the values of the dc transfer functions:
$H_{NE}(0)$ and $H_{FE}(0)$. These are computed (either by hand or with PSPICE)
from Fig. 6.11 as $H_{NE}(0) = 1.618E - 3$ and $H_{FE}(0) = -1.618E - 3$. The dc
resistance of each land is computed as

$$r_{dc}\,\mathscr{L} = \frac{\mathscr{L}}{\sigma w t}$$
$$= 0.3279\ \Omega$$

Figure 6.12(a) and (b) compare the predicted near- and far-end crosstalk
voltages computed (a) neglecting losses with the SPICEMTL.FOR subcircuit
model and (b) incorporating frequency-dependent lossy conductor losses

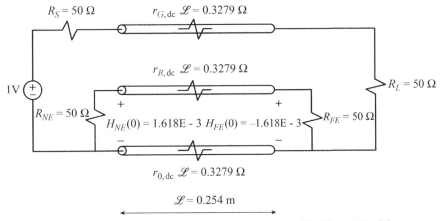

FIGURE 6.11. Computing the dc transfer functions $H_{NE}(0)$ and $H_{FE}(0)$.

with MTL.FOR and the resulting time-domain waveforms computed with TIMEFREQ.FOR (TIMEFREQ.EXE).

Observe that the conductor losses cause a shift in levels of 1.6 mV caused by the common-impedance coupling. But other than that, the conductor losses have little effect on the crosstalk waveforms $V_{NE}(t)$ and $V_{FE}(t)$. Again, the magnitudes of the frequency-domain transfer functions in Fig. 6.6(a) and Fig. 6.7(a) show a level of $|\hat{H}(0)| = 1.6 \times 10^{-3}$ at low frequencies caused by the common-impedance coupling, while the angle approaches $\angle\hat{H}(0) \rightarrow 0, \pm 180°$.

The conductor losses for this example don't seem to have a significant effect on the crosstalk waveforms here except when $V_S(t)$ is in its steady-state value of 1 V. This is due to the dominance of the 50-Ω terminal resistances dominating the line resistances of some $0.3279\,\Omega$. An example of where conductor losses can have a significant effect on the crosstalk waveform shape for a line having significant conductor losses is shown in

C. R. Paul, *Analysis of Multiconductor Transmission Lines*, 2nd ed., Wiley-Interscience, Hoboken, NJ, 2008, pp. 439–447.

The experimentally obtained results for the near-end crosstalk are shown in Fig. 6.13 and match the predictions in Fig. 6.12(a) reasonably well. The digital oscilloscope used to obtain these experimental results has a bandwidth of 500 MHz, whereas the bandwidth of the source voltage is

$$\text{BW} = \frac{1}{\tau_r} = \frac{1}{1.25\ \text{ns}} = 800\ \text{MHz}.$$

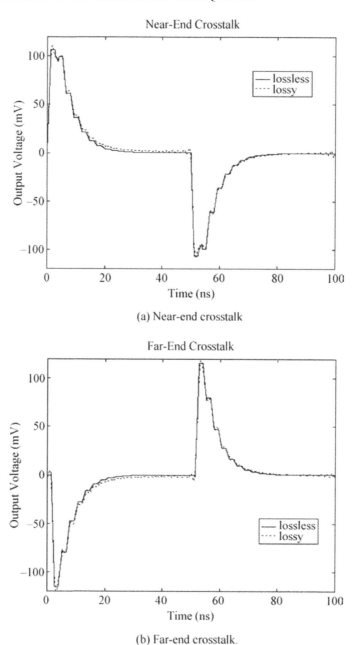

(a) Near-end crosstalk

(b) Far-end crosstalk.

FIGURE 6.12. (a) Near- and (b) far-end crosstalk for the circuit in Fig. 6.5 with $\tau_r = \tau_f = 1.25$ ns computed neglecting losses using the subcircuit model generated by SPI-CEMTL.FOR compared to the crosstalk including the losses of the conductors computed by MTL.FOR and TIMEFREQ.FOR.

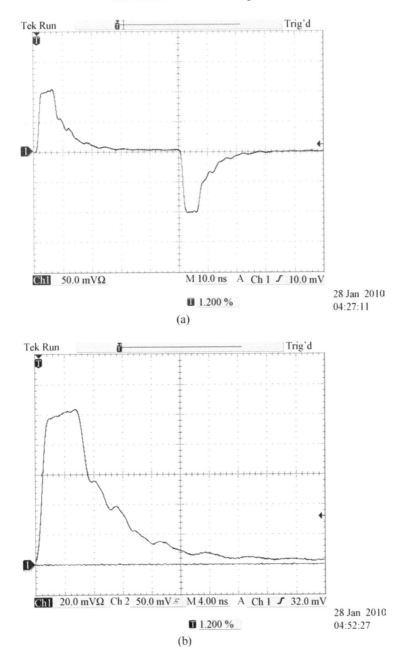

FIGURE 6.13. Results obtained experimentally for the near-end crosstalk compared to the predictions in Fig. 6.12(a). Figure 6.13(a) shows the near-end predicted crosstalk voltage, and Figs. 6.13(b) and 6.13(c) show close-up views.

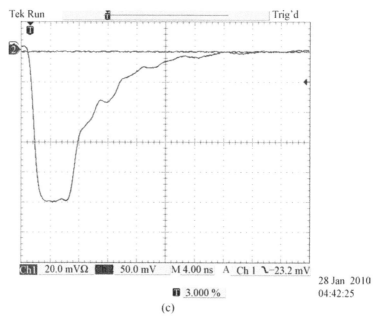

(c)

FIGURE 6.13 (*Continued*)

PROBLEMS

6.1 Derive the MTL equations for a lossy line in (6.9).

6.2 Show, by direct substitution, that (6.48) satisfies (6.9).

6.3 Compare the lossless and lossy crosstalk time-domain waveforms for the circuit in Fig. P6.3. Plot the magnitude and phase of the frequency-domain transfer functions with and without losses.

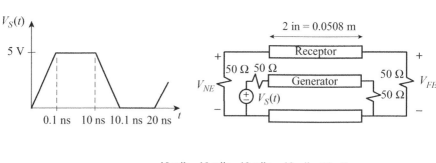

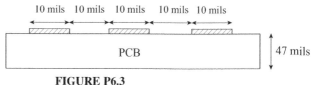

FIGURE P6.3

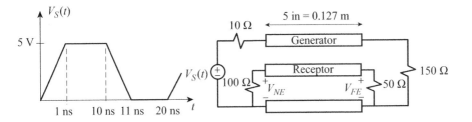

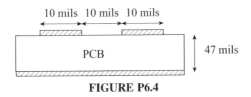

FIGURE P6.4

6.4 Compare the lossless and lossy crosstalk time-domain waveforms for the circuit in Fig. P6.4. Omit the loss in the ground plane. Plot the magnitude and phase of the frequency-domain transfer functions with and without losses.

APPENDIX

A BRIEF TUTORIAL ON USING PSPICE

This is a brief summary of SPICE (simulation program with integrated-circuit emphasis) and its personal computer version PSPICE, electric circuit analysis program. The original SPICE computer program was developed to analyze complex electric circuits, particularly integrated circuits. It was developed at the University of California at Berkeley in the late 1960s. Since it was developed under U.S. government funding, it is not proprietary and can be freely copied, used, and distributed. This was written for use on large mainframe computers of the time. In the early 1980s the MicroSim Corporation developed a personal computer version of SPICE called PSPICE. A number of important modifications were made, particularly in the plotting of data via the . PROBE function. Since then a number of commercial firms have modified and developed their own PC versions. But essentially, the core engine is that of the original SPICE code. The MicroSim version of PSPICE, version 8, was acquired by Cadence Design Systems. Theirs is version 10.0, called OrCAD Capture, which also contains the primary simulation code PSPICE A/D. A Windows-based version is available free from www.cadence.com. The OrCAD Capture program was originally called Schematic in the MicroSim version and contains a number of enhancements. A number of books [1–5] detail the use of SPICE and PSPICE. Both the MicroSim version 8 and the OrCAD version 10 are contained in a CD at the end of this and other textbooks [6].

Transmission Lines in Digital and Analog Electronic Systems: Signal Integrity and Crosstalk, By Clayton R. Paul
Copyright © 2010 John Wiley & Sons, Inc.

There are two methods of entering and executing a PSPICE program. The first method is the *direct method* described here, where one enters the program code using an ASCII text editor (supplied with PSPICE). Note: SPICE and PSPICE make no distinction between lowercase and uppercase letters. Then this text file is run using the PSpice A/D section of the program, and the output is examined with the text editor. The second method is the *schematic method* (now called *capture*), where the user "draws" the circuit diagram directly on the screen and then executes that program. The direct method is generally the most rapid method of solving relatively simple problems. The schematic (capture) method has the advantage of visually seeing whether the circuit components are connected as intended for more complex circuits but for most simple EMC problems is a bit more time consuming to set up than the direct method since numerous windows and drop-down menus must be navigated.

Once the PSPICE program has been installed on your computer, the following is a description of how you can input your program, run it, and examine the output. The various selections are underlined. Although there are several ways of doing this, the simplest is to use the Design Manager. To load this, you click or select the following in this sequence.

1. Start
2. Programs
3. MicroSim Eval 8
4. Design Manager

The direct method is simply to type in the PSPICE program using the TextEdit feature. To enter this and prepare the program, we select the following in this sequence.

1. Select TextEdit (lower button on the vertical toolbar on the left).
2. Type the program.
3. Save the program as XXX.cir or XXX.in and close it.
4. Select PSpice A/D (second button on the vertical toolbar on the left).
5. Click on File, Open and select the file stored previously. The program will run automatically and the output will be stored in file XXX.out.
6. Click on File, Run Probe to plot waveforms or File, Examine Output to examine the printed output.
7. Alternatively, you could recall the TextEdit progam and select File, Open, XXX.out to examine the printed output, which is self-explanatory.

CREATING THE SPICE OR PSPICE PROGRAM

SPICE and PSPICE write the node-voltage equations of an electric circuit [1]. One node, the reference node for the node voltages, is designated the zero (0) node. All circuits must contain a zero node. The other nodes in the circuit are labeled with numbers or letters. For example, a node may be labeled 23 or it may be labeled FRED. The voltages with respect to the reference node are positive at the node and denoted as V(N1), V(N2), and so on, as shown in Fig. A.1.

The general structure of any SPICE or PSPICE program is as follows:

1. Title
2. { Circuit Description
3. Execution Statements
4. Output Statements
5. .END

The first line of the SPICE program is the Title and is not processed by SPICE. It is simply written on the output and any plots. A comment line is started with an asterisk (*) and also is not processed by the program. A line may be continued with a plus sign (+) at the beginning of the following line. The next set of lines, Circuit Description, describes the circuit elements and their values and tells SPICE how they are connected together to form the circuit. The next set of lines are the Execution Statements that tell SPICE what type of

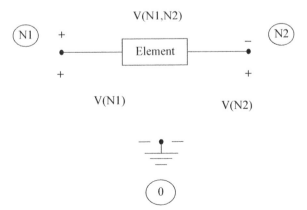

FIGURE A.1. Node voltage and element voltage notation in the SPICE (PSPICE) circuit analysis program.

analysis is to be run: dc analysis (.DC), sinusoidal steady state or phasor analysis (.AC), or the full time-domain analysis, consisting of the transient and steady-state solution (.TRAN). The next set of statements, Output Statements, tell SPICE what outputs are desired. The results can be printed to a file with the .PRINT statement or can be plotted with the .PROBE feature. Finally, all programs must end with the .END statement. Actually, items 2 to 4 above can appear in any order in the program, but the program must begin with a Title statement and end with the .END statement.

CIRCUIT DESCRIPTION

The basic elements and their SPICE descriptions are shown in Fig. A.2. Figure A.2(a) shows the independent voltage source. It is named starting with the letter V and then any other letters. For example, a voltage source might be called VFRED. It is connected between nodes N1 and N2. It is very important to note that *the source is assumed positive at the first-named node*. The current through the voltage source is designated as I(VXXX) and is assumed to flow from the first-named node to the last-named node. The source type can be either dc, for which we append the terms DC *magnitude*, or a sinusoid, to which we append the terms AC *magnitude phase* (*degrees*). A time-domain waveform is described by several functions which we describe later, and these descriptions are appended (without the word TRAN). The independent current source is shown in Fig. A.2(b). Its name starts with the letter I followed by any other letters. For example, a current source might be designated as ISAD. *The current of the source is assumed to flow from the first-named node to the last-named node.* The types of source waveforms are the same as for the voltage source.

The resistor is shown in Fig. A.2(c) and its name starts with the letter R (e.g., RHAPPY). The current through the resistor is designated as I(RXXX) and is assumed to flow from the first-named node to the last-named node. SPICE does not allow elements with zero values. Hence a resistor whose value is $0\,\Omega$ (a short circuit) may be represented as having a value of 1E-8 or any other suitably small value. Similarly, an open circuit may be designated as a resistor having a value of 1E8 or any other suitably large value. Every node must have at least two elements connected to it. Also, SPICE requires that every node must have a dc path to ground [the zero (0) node]. Placing a large resistor (e.g., 1MEG Ohm) between such nodes fixes this problem.

The inductor is shown in Fig. A.2(d) and is designated with the letter L (e.g., LTOM). The current through the inductor as well as the initial inductor current

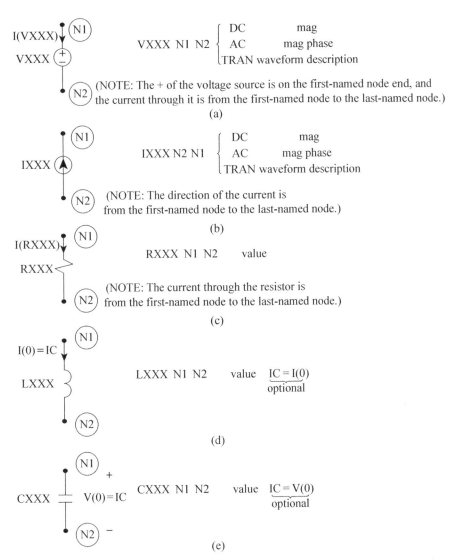

FIGURE A.2. Coding convention for (a) the independent voltage source, (b) the independent current source, (c) the resistor, (d) the inductor, and (e) the capacitor.

at $t = 0^+$, I(0), is assumed to flow from the first-named node to the last-named node. The initial condition can be specified at the end of the statement with IC = I(0). The capacitor is shown in Fig. A.2(e) and is designated with the letter C (e.g., CME). The initial voltage across the capacitor at $t = 0^+$, V(0), can be specified at the end of the statement with IC = V(0), and this voltage is assumed to be positive at the first-named node.

All numerical values can be specified in powers of 10 and written in exponential format (e.g., $2 \times 10^{-5} = $ 2E-5) or by using standard multipliers using standard engineering notation:

Multiplier	SPICE Symbol
10^9 (giga)	G
10^6 (mega)	MEG
10^3 (kilo)	K
10^{-3} (milli)	M
10^{-6} (micro)	U
10^{-9} (nano)	N
10^{-12} (pico)	P

For example, 1 megohm is written as 1MEG, 1 kilohm is written as 1K, 3 millihenries is written as 3M, 5 microfarads is written as 5U, 2 nanohenries is written as 2N, and 7 picofarads is written as 7P. A 3-farad capacitor should not be written as 3F, since F stands for femto $= (10^{-15})$.

The four types of controlled sources, G, E, F, H, are shown in Fig. A.3 along with their descriptions. The polarities of voltage and the currents through the elements conform to the previous rules governing these in terms of the first- and last-named nodes on their description statements. For a current-controlled source, F or H, the controlling current must be through an independent voltage source. Often we insert a 0-V source to sample the current. Some recent versions of PSPICE allow the specification of the current through any element as a controlling current. But it is always a simple matter to insert a zero-volt voltage source.

Figure A.4 shows how to specify mutual inductance. First, the self-inductances are coupled and specified as before. The mutual inductance is specified in terms of its coupling coefficient:

$$k = \frac{M}{\sqrt{L_1 L_2}}$$

To keep the polarities correct, define the self-inductances so that the dots are on the first-named nodes when the two inductors are defined; otherwise, a negative coupling coefficient may need to be specified.

Figure A.5 shows the last important element, the transmission line (lossless), which is used extensively in signal integrity analyses. This gives the *exact solution* of the transmission-line equations for a *lossless* line. There are many ways to specify the important parameters for the (lossless) line, but the one shown in the figure is the most widely used: Specify the characteristic impedance of the line and the line's one-way time delay. Alternatively, you can create a lumped RLCG *approximate* model of a *lossy* line using lumped-circuit elements [6]. But the per-unit-length parameters of resistance R and

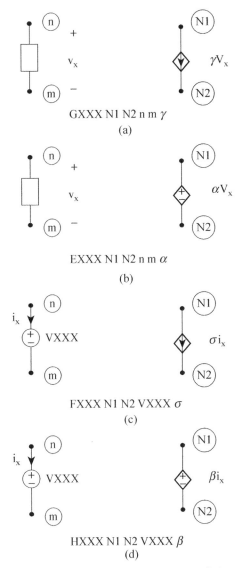

GXXX N1 N2 n m γ

(a)

EXXX N1 N2 n m α

(b)

FXXX N1 N2 VXXX σ

(c)

HXXX N1 N2 VXXX β

(d)

FIGURE A.3. Coding convention for (a) the voltage-controlled current source, (b) the voltage-controlled voltage source, (c) the current-controlled current source, and (d) the current-controlled voltage source.

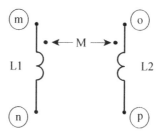

L1 m n value
L2 o p value
KXXX L1 L2 coefficient coupling

$$\text{coupling coefficient } \frac{M}{\sqrt{L1\,L2}}$$

FIGURE A.4. Coding convention for mutual inductance between two coupled inductors.

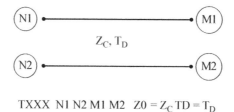

TXXX N1 N2 M1 M2 $Z0 = Z_C$ $TD = T_D$

FIGURE A.5. Coding convention for the two-conductor lossless transmission line.

conductance G in that model must be constants, although these are, in reality, frequency dependent, which is not readily handled in an exact solution of the transmission-line equations [6].

Figure A.6 shows how to specify the important time-domain waveforms. Figure A.6(a) shows the PWL (piecewise-linear) waveform, where straight lines are drawn between pairs of points that are specified by their time location and their value. Observe that the function holds the last value specified, V4 in the figure. Figure A.6(b) shows the periodic pulse waveform, PULSE, that is used to specify a periodic clock or other timing waveforms. The function specifies a trapezoidal waveform that repeats periodically with period PER (the reciprocal is the fundamental frequency of the waveform). Note that the pulse width, PW, is not specified between the 50% points of the pulse, as is the usual convention.

The sinusoidal function is specified by

```
SIN(Vo Va [[Freq [[Td [[Df [[Phase]]]]]]]])
```

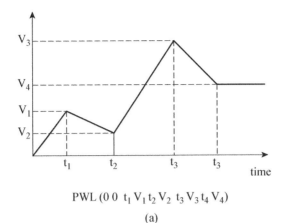

$$\text{PWL} (0\ 0\ t_1\ V_1\ t_2\ V_2\ t_3\ V_3\ t_4\ V_4)$$

(a)

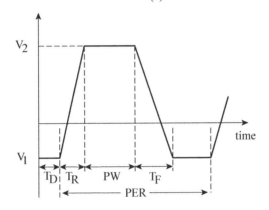

$$\text{PULSE } V_1\ V_2\ T_D\ T_R\ T_F\ \text{PW PER}$$

(b)

FIGURE A.6. Coding convention for the important source waveforms: (a) the piecewise-linear waveform; (b) the pulse source waveform (periodic).

which gives the waveform

$$x(t) = \text{Vo} + \text{Va} \sin\left(2\pi\left(\text{Freq}(\text{time} - \text{Td}) + \frac{\text{Phase}}{360}\right)\right) e^{-(\text{time} - \text{Td})\text{Df}}$$

Brackets around items signify that they are optional. Hence to specify the general sinusoidal waveform

$$x(t) = A\sin(n\omega_0 t + \theta)$$

where $\omega_0 = 2\pi f_0$ and $f_0 = 1/\text{PER}$ is the frequency of the sinusoid we would write

$$\text{SIN}(0\ A\ n f_0\ 0\ 0\ \theta)$$

EXECUTION STATEMENTS

There are three types of solutions: dc, sinusoidal steady state or phasor, and the full time-domain solution (so-called transient although it contains both the transient and the steady-state parts of the solution).

The dc solution is specified by

.DC V,IXXX *start_value end_value increment*

where V,IXXX is the name of a dc voltage or current source in the circuit whose value is to swept. For example, to sweep the value of a dc voltage source VFRED from 1 V to 10 V in increments of 2 V and solve the circuit for each of these source values, we would write

.DC VFRED 1 10 2

If no sweeping of any source is desired, simply choose one dc source in the circuit and iterate its value from the actual value to the actual value and use any nonzero increment. For example,

.DC VFRED 5 5 1

The sinusoidal steady-state or phasor solution is specified by

.AC{LIN,DEC,OCT} *points start_frequency end_frequency*

LIN denotes a linear frequency sweep from *start_frequency* to *end_frequency*, and *points* is the total number of frequency points. DEC denotes a log sweep of the frequency, where the frequency is swept logarithmically from the *start_frequency* to the *end_frequency*, and *points* is the number of frequency points per decade. OCT is a log sweep by octaves, where *points* is the number of frequency points per octave.

The time-domain solution is obtained by specifying

.TRAN *print_step end_time [no_print_time]*
       [step_ceiling] [UIC]

SPICE solves the time-domain differential equations of the circuit by discretizing the time variable into increments of Δt and solving the equations in a bootstrapping manner. The differential equations of the circuit are first solved at $t = 0$. Then that solution is used to give the solution at $t = \Delta t$. These prior time solutions are then used to give the solution at $t = 2\Delta t$, and so on. The first item, *print_step*, governs when an output is requested. Suppose that the discretization used in the solution is every 2 ms. We might not want to see (in the output generated by the .PRINT statement) an output at every 2 ms but only every 5 ms. Hence we might set the *print_step* time as 5M. The *end_time* is the final time for which the solution is obtained. The remaining parameters are optional. The analysis always starts at $t = 0$. But we may not wish to see a printout of the solution (in the output generated by the .PRINT statement) until after some time has elapsed. If so, we would set the *no_print_time* to that starting time. SPICE and PSPICE have a very sophisticated algorithm for determining the minimum step size, Δt, for discretization of the differential equations in order to get a valid solution. The default maximum step size is *end_time*/50. However, there are many cases where we want the step size to be smaller than what SPICE would allow, in order to increase the accuracy of the solution or to increase the resolution of the solution waveforms. This is frequently the case when we use SPICE in the analysis of transmission lines where abrupt waveform transitions are occurring at every one-way time delay. The *step_ceiling* is the maximum time step size, $\Delta t|_{max}$, that will be used in discretizing the differential equations, as described previously. Although this gives longer run times, there are cases where we need to do this to generate the required accuracy. The last item, *UIC*, means that SPICE is to use the initial capacitor voltage or inductor current specified on the element specifications with the *IC* = command. In a transient analysis, SPICE will compute the initial conditions at $t = 0^+$. If some other initial conditions are required, we should set these on the capacitor or inductor specifications with the *IC* = command and specify *UIC* on the .TRAN statement. For example,

```
.TRAN 0.5N 20N 0 0.01N
```

would command SPICE to do a full time-domain (transient plus steady state) analysis for times from 0 to 20 ns, print out a solution at every 0.5 ns, start printing to the output file at $t = 0$, and use a time-discretization time step no larger than 0.01 ns.

OUTPUT STATEMENTS

The output statements are either for printing to a file with the .PRINT statement or producing a plotted graph of any waveform with the .PROBE statement. The .PRINT statement has three forms, depending on the type of analysis being run. For a dc analysis,

```
.PRINT DC V(X) I(R)
```

prints the dc solution for the voltage of node X with respect to the reference node, and I(R) prints the dc solution for current through resistor R (defined from the first-named node to the last-named node on the specification statement for resistor R). For a sinusoidal steady-state analysis (phasor solution),

.PRINT AC VM(NI) VP(NI) IM(RFRED) IP(RFRED)

prints the magnitude and phase of node voltages and currents where the magnitude and phase of the node voltage at node NI are VM(NI) and VP(NI), respectively. For the currents through a resistor RFRED, the magnitude is IM(RFRED) and the phase is IP(RFRED). For the time-domain or transient analysis, the print statement

.PRINT TRAN V(NI) I(RFRED)

prints the solutions at all solution time points (specified on the .TRAN line) for the voltage at node NI with respect to the reference node, and the current through resistor RFRED (defined from the first-named node to the last-named node on the specification statement for resistor RFRED).

In addition, the .FOUR statement computes the expansion coefficients for the Fourier series (magnitude and phase), $c_n \angle c_n$:

.FOUR f_0 [output_variable(s)]

The .FOUR command can only be used in a .TRAN analysis. The fundamental frequency of the *periodic* waveform to be analyzed is denoted as $f_0 = 1/T$, where T is the period of the waveform. The *output_variable(s)* are the desired voltage or current waveforms to be analyzed [e.g., V(2), I(R1)]. The phase results are with reference to a sine form of the series

$$x(t) = c_o + \sum_{n=1}^{\infty} c_n \sin(n\omega_0 t + \angle c_n)$$

There is an important consideration in using the .FOUR command. The portion of the waveform that is analyzed to give the Fourier expansion coefficients is the last portion of the solution time of length one period, $1/f_0 = T$. In other words, SPICE determines the coefficients from the waveform between $end\_ime - [1/f_0]$ and end_time. Hence, end_time on the . TRAN command should be at least one period long. In situations where the solution has a transient portion at the beginning of the solution interval and we want to determine the Fourier coefficients for the steady-state solution, we would run the analysis for several periods to ensure that the solution has gotten into steady state. For example, consider an input signal that is periodic with a period of 2 ns or a fundamental frequency of 500 MHz. An output voltage at node 4 would also have this periodicity but would have a transient period of at least five time constants. Suppose that the maximum time constant of the circuit is 4 ns. Then we would set the end time to 20 ns or more in order to get into the steady-state region of the solution at the end of the solution. The following commands would be used to obtain the Fourier coefficients of the steady-state response of the node voltage at node 4:

```
.TRAN 0.1N 20N
.FOUR 500MEG V(4)
```

This would compute the solution for the voltage waveform at node 4 from $t = 0$ to $t = 20$ ns. Since the period (the inverse of 500 MHz) is specified as 2 ns, the portion of the waveform from 18 to 20 ns would be used to compute the Fourier coefficients for the waveform. If we wanted to compute the Fourier coefficients for the initial part of the waveform, including the transient, we would specify

```
.TRAN 0.1N 2N
```

which would run for only one period. There is a FFT button on the toolbar that can compute the fast Fourier transform of the waveform. Essentially, this valuable feature can turn PSPICE into a "poor man's spectrum analyzer."

All printed output is directed to a file named XXXX.OUT if the input file is named XXXX.IN or XXXX.CIR. Plotting waveforms is the greatest enhancement of PSPICE over the original SPICE. This is invoked simply by placing the .PROBE statement in the program list. No additional parameters are required. PSPICE stores all variables at all solution time points and waits for the user to specify which to plot.

EXAMPLES

In this brief tutorial we have shown the basic commands that one can use to solve the vast majority of electric circuit analysis problems, particularly those that are encountered in EMC problems. Essentially, PSPICE performs the tedious and laborious solutions of lumped-circuit models of the problem. Solving these circuits by hand would not only be time consuming but would involve so many errors as to make the solutions obtained useless. To simplify the learning, we have conscientiously minimized the detail and purposely not shown all the possible options. However, there are a myriad of options that can simplify many computations, and the reader should consult the references.

EXAMPLE 1

Use PSPICE to compute the voltage V_{out} and the current I in the circuit of Fig. A.7.

Solution The PSPICE coding diagram with nodes numbered is shown in Fig. A.7. Zero-volt voltage sources are inserted to sample the currents i_x and i. The PSPICE program is

```
EXAMPLE 1
VS 1 0 DC 5
R1 1 2 500
R2 2 3 1K
R3 3 4 2K
VTEST1 4 0 DC 0
HSOURCE 3 5 VTEST1 500
```

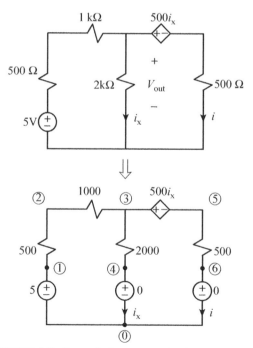

FIGURE A.7. Example 1, an example of a DC analysis.

```
R4 5 6 500
VTEST2 6 0 DC 0
.DC VS 5 5 1
*THE CURRENT I IS I(VTEST2) AND THE VOLTAGE VOUT IS V(3)
 + OR V(3,4)
.PRINT DC V(3) I(VTEST2)
.END
```

The result is $i = I(\text{VTEST2}) = 1.875\text{E-}3$ and the voltage Vout $= V(3) = 1.250\text{E0}$.

EXAMPLE 2

Use PSPICE to plot the frequency response of the bandpass filter shown in Fig. A.8(a).

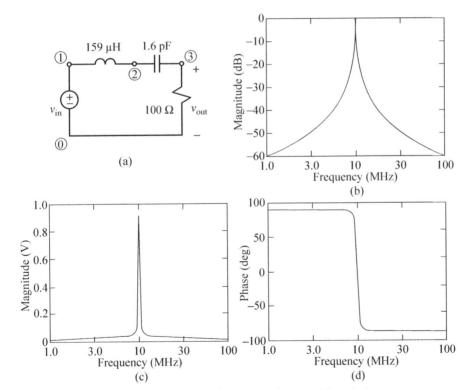

FIGURE A.8. Example 2, an example of an AC analysis.

Solution   The nodes are numbered on the circuit diagram and the PSPICE program is

```
EXAMPLE 2
VS 1 0 AC 1 0
RES 3 0 100
LIND 1 2 159U
CAP 2 3 1.6P
.AC DEC 50 1MEG 100MEG
.PROBE
*THE MAGNITUDE OF THE OUTPUT IS VM(3) AND THE PHASE IS
 + VP(3)
.END
```

The magnitude of the voltage is plotted in Fig. A.8(b) in decibels using VDB (3), which means

$$VDB(3) = 20\log_{10}VM(3)$$

Figure A.8(c) shows what we get if we request VM(3): The data are highly compressed outside the bandpass region. The phase (in degrees) is plotted in Fig. A.8(d). The resonant frequency is 10 MHz. The phase is $+90°$ below the resonant frequency, due to the dominance of the capacitor in this range, and is $-90°$ above the resonant frequency, due to the dominance of the inductor in this range. This bears out the important behavior of a series resonant circuit.

EXAMPLE 3

Use PSPICE to plot the inductor current for $t > 0$ in the circuit of Fig. A.9(a).

Solution   The circuit immediately before the switch opens (i.e., at $t = 0^-$) is shown in Fig. A.9(b), from which we compute the initial voltage of the capacitor as 4 V and the initial current of the inductor as 2 mA. The PSPICE diagram with nodes numbered is shown in Fig. A.9(c), and the PSPICE program is

```
EXAMPLE 3
IS 0 1 DC 10M
R 1 2 2K
```

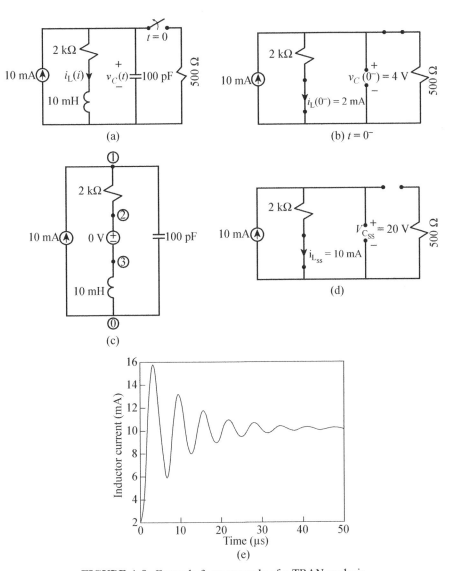

FIGURE A.9. Example 3, an example of a TRAN analysis.

```
VTEST 2 3
L 3 0 10M IC = 2M
C 1 0 100P IC = 4
.TRAN .05U 50U 0 .05U UIC
*THE INDUCTOR CURRENT IS I(VTEST) OR I(L)
.PROBE
.END
```

We have chosen to solve the circuit out to $50\,\mu s$, print the solution in steps of $0.05\,\mu s$, and have directed PSPICE to use a solution time step no larger that $0.05\,\mu s$ as well as to use the initial conditions given for the inductor and capacitor. The result is plotted using PROBE in Fig. A.9(e). The result starts at $2\,mA$, the initial inductor current, and eventually converges to the steady-state value of $10\,mA$, which can be confirmed by replacing the inductor with a short circuit and the capacitor with an open circuit in the $t > 0$ circuit, as shown in Fig. A.9(d).

EXAMPLE 4

Figure A.10 shows an example where an interconnecting set of conductors (lands on a PCB) can cause severe logic errors, resulting in poor signal integrity. Two CMOS inverters (buffers) are connected by 2 in of lands ($\mathscr{L} = 2\,in = 5.08\,cm$) on a PCB. The output of the left inverter is shown as a Thevenin equivalent circuit having a low source resistance of $10\,\Omega$. This is fairly typical of CMOS devices except that the output resistance is somewhat nonlinear. The load on the line is the input to the other CMOS inverter, which is represented as a 5-pF capacitor that is also typical of the input to CMOS devices. We are interested in determining the voltage at the output of the interconnect line, $V_L(t)$, which is the voltage at the input to the second CMOS inverter. The open-circuit voltage of the left inverter, $V_S(t)$, is a 5-V 50-MHZ (period of 20 ns), 50% duty cycle clock trapezoidal waveform having 0.5-ns rise and fall times. A 10-mil land lying on a glass-epoxy PCB of 47 mils thickness has a ground plane below it. For the cross-sectional dimensions of the line the characteristic impedance is $Z_C = 124\,\Omega$, and the velocity of propagation is $v = 1.7 \times 10^8\,m/s$. This gives a one-way time delay of

$$T_D = \frac{\mathscr{L}}{v} = \frac{0.0508\,m}{1.7 \times 10^8\,m/s} = 0.3\,ns$$

The response for the load voltage is computed with PSPICE using the exact (lossless) transmission-line model contained in PSPICE and is shown in Fig. A.10(b).

Solution   The nodes are numbered on the circuit diagram, and the PSPICE program is

```
EXAMPLE 4
VS 1 0 PULSE (0 5 0 0.5N 0.5N 9.5N 20N)
RS 1 2 10
```

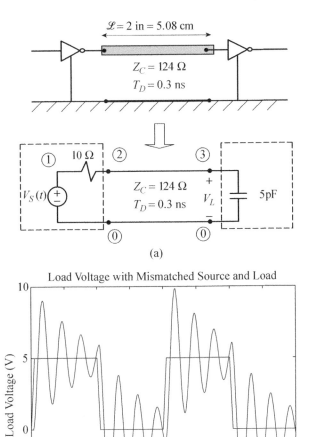

FIGURE A.10. Example 4, an example of signal integrity analysis.

```
T 2 0 3 0 Z0=124 TD=0.3N
CAP 3 0 5P
.TRAN 0.05N 40N 0 0.05N
.PROBE
.END
```

Typical thresholds for CMOS circuits are approximately halfway between the logic 1 and logic 0 levels, which in this case are 5 V and 0 V. Observe that there is severe ringing in the response (the input to the second inverter) and the response drops below the 2.5 V high level and rises above the 2.5 V low level, thereby producing false logic triggering. Hence signal integrity is not achieved here.

THE SUBCIRCUIT MODEL

SPICE (PSPICE) has a handy way of utilizing models of devices in several places in a SPICE program without having to redefine these models at every place of usage. This is similar to the subroutine in FORTRAN. It is called the SUBCKT model. For example, suppose that we have developed an extensive model of, say, an operational amplifier or a multiconductor transmission line (MTL) [6]. The subcircuit model might have, for example, four external nodes that we have named 101, 102, 201, 202, as illustrated in Fig. A.11 for a subcircuit model of a MTL.

The nodes internal to the model are unique to this model and have no resemblance to the nodes of the SPICE program into which this model is to be embedded (perhaps at several locations). However, the zero (0) or universal ground node is the only node that is common with the main program. The subcircuit model description is

$$\text{.SUBCKT MTL 101 102 201 202}$$

$$\cdots \cdots \cdots$$

$$\cdots \cdots \cdots$$

$$\text{.ENDS MTL}$$

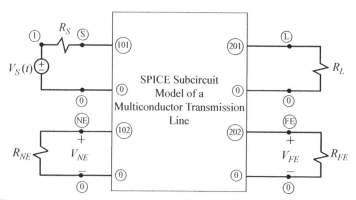

FIGURE A.11. Connection of a SPICE subcircuit model of a multiconductor transmission line (MTL) to its terminations.

where MTL is the name given to the subcircuit, and its external nodes are 101, 102, 201, and 202. The subcircuit model ends with .ENDS and the name of the subcircuit model, MTL. These node numbers are unique to the subcircuit, *but their ordering is important.* The call statement in the SPICE main program is formatted as

```
XMTL S NE L FE MTL
```

Hence the nodes of the subcircuit MTL are attached to the external nodes of the main SPICE model as S = 101, NE = 102, L = 201, FE = 202 to connect the subcircuit to the MTL terminations, as illustrated in Fig. A.11. The subcircuit model must end with

```
.ENDS MTL
```

and the main SPICE program must end with the statement

```
.ENDS
```

REFERENCES

1. C. R. Paul, *Fundamentals of Electric Circuit Analysis*,Wiley, New York, 2001.
2. P. W. Tuinenga, *SPICE: A Guide to Simulation and Analysis Using PSPICE*, 3rd ed., Prentice Hall, Englewood Cliffs, NJ, 1995.
3. A. Vladimirescu, *The SPICE Book*,Wiley, New York, 1994.
4. R. Conant, *Engineering Circuit Analysis with PSpice and Probe*, McGraw-Hill, New York, 1993.
5. J. W. Nilsson and S. A. Riedel, *Introduction to PSpice Manual for Electric Circuits Using OrCad Release 9.1*, 4th ed., Prentice Hall, Upper Saddle River, NJ, 2000.
6. C. R. Paul, *Introduction to Electromagnetic Compatibility*, 2nd ed.,Wiley-Interscience, Hoboken, NJ, 2006.

INDEX

"effective" relative permittivity, 43
 microstrip line, 69
 PCB, 43
.AC
 in SPICE, 276
.DC
 in SPICE, 276
.TRAN
 in SPICE, 276

alternative results for the line voltage and
 current
 phasor solution, 137
Ampere's law, 58

bandwidth (BW) of trapezoidal
 waveform, 22, 26
bandwidth and signal integrity, 104
bandwidth of digital signals, 17, 19
bound dipoles, 150
 in ribbon cables, 197
bounds and envelope of digital
 waveform, 16, 26

bounds of digital waveform, 16
bounds of Fourier coefficients
 of trapezoidal pulse, 14

capacitors
 in SPICE, 276, 277
characteristic impedance, 56
 lossy line, 151
CMOS thresholds, 34
common-impedance coupling, 260
conductor losses
 two-conductor lines, 149
construction of microwave circuit
 components
 with transmission lines, 167
controlled sources
 in SPICE, 278, 279
coupled inductors
 in SPICE, 278, 280
coupled phasor transmission-line equations
 two-conductor lines, 122
crosstalk, 41
cyclic frequency, f, 4

*Transmission Lines in Digital and Analog Electronic Systems: Signal Integrity
and Crosstalk,* By Clayton R. Paul
Copyright © 2010 John Wiley & Sons, Inc.

decoupling the MTL equations
  homogeneous media, 213
  inhomogeneous media, 214
dielectric losses
  two-conductor lines, 149
driving multiple lines, 111
duty cycle
  of trapezoidal pulse, 13

effect of reactive terminations
  capacitive, 92
  inductive, 94
effects of line losses
  two-conductor lines, 147
electrical dimensions
  in terms of time delay, 9
electrical lengths, 2
electrically short dimensions
  in terms of time delay, 10, 11
embedding a linear system, 31
English system of units, 5
envelope of Fourier coefficients
  of trapezoidal pulse, 14

fall times, 1
far-end crosstalk, 41
forward-traveling and
    backward-traveling waves, 56
Fourier series, 12
Fourier transform, 13
frequency-domain of two-conductor
    lines
  for sinusoidal steady-state phasor
    excitation, 122
frequency-domain transfer
    function, 45

Gauss's law, 59
general phasor solution
  two-conductor lines, 125
generator circuit, 41
  three-conductor lines, 178
harmonics, 12
homogeneous medium, 55

ideal load voltage, 34
independent current sources
  in SPICE, 276, 277
independent voltage sources

in SPICE, 276, 277
inductive-capacitive approximate model, 232
inductors
  in SPICE, 276, 277
input impedance to line
  phasor solution, 123, 125
Internation System (SI), 3

Kirchhoff's Laws, 10

lands, 1
Laplace transform
  solution of transmission-line equations, 93
line discontinuities
  effect of, 105
linearity of engineering systems, 27
lossless line, 52
lumped-circuit approximate models of line,
    36, 91
lumped-circuit approximations, 161
lumped-circuit model of crosstalk, 43
lumped-circuits
  in terms of time delay, 11

matching and VSWR, 133
MATLAB, 20, 22
mean absolute error (MAE), 18
mean-square error (MSE), 17
method of images, 64
mils, 5
MKSA system, 3
MSTRP.FOR,MSTRP.EXE, 203
MTL conductor losses, 245
MTL dielectric losses
  homogeneous medium, 248, 249
  inhomogeneous media, 250
MTL.FOR,MTL.EXE
  computer program, 254
MTL.IN, 256
MTL.OUT, 264
MTLFREQ.DAT, 266
multiconductor transmission line equations
  coupled, 181
  coupled, phasor, 183
  homogeneous medium, 182
  uncoupled, 182
  uncoupled, phasor, 183
multipliers
  in SPICE, 278

near-end crosstalk, 41
node voltages
  in SPICE, 275
numerical methods
  for computing the per-unit-length
    parameter matrices, 196

Parseval's theorem, 18
PCB.FOR, PCB.EXE
  computer programs, 201
PCBs, 200
periodic waveform, 12
per-length capacitance
  one wire above ground, 65
per-length inductance
  one wire above ground, 65
permeability
  of free space, 3
permittivity
  of free space, 3
per-unit-length capacitance, 52
  coaxial cable, 67
  microstrip line, 69
  PCB, 70
  stripline, 68
  two wires, 63
per-unit-length inductance, 52
  coaxial cable, 67
  mcrostrip line, 69
  PCB, 70
  stripline, 68
  two wires, 63
per-unit-length matrices
  lossy MTLs, 243
per-unit-length parameters
  multiconductor lines, 184
  two-conductor wire-type line, 57
per-unt-length capacitance
  PCB, 70
phase constant, β, 6
phasor transfer function, 29, 34
phasor transmission lines
  general solution, 123, 124
plane waves, 11, 52
pointwise error, 17
power flow on the line, 134
proximity effect, 64
PUL.DAT, 192, 201, 203, 204, 219
pulse width, 12

radian frequency, ω, 4
receptor circuit, 41
  three-conductor lines, 178
resistors
  in SPICE, 276, 277
ribbon cables, 197
RIBBON.FOR, RIBBON.EXE
  computer programs, 199
rise times, 1

series solution, 83
signal integrity, 2, 32
  matching at the source and at the
    load, 103
  matching schemes, 96
  overshoot and undershoot, 96
  parallel matching, 98
  series matching, 97
similarity transformations
  to decouple losslees MTLs, 209
$\sin(x)/x$, 13
skin effect
  two-conductor lines, 149
small-loss approximations, 151
Smith chart, 138
solution of MTL transmission-line equations
  lossy lines, 251
speed of light
  in free space, 4
SPICE (PSPICE) solution, 84, 128
SPICE subcircuit model (SUBCKT), 217
  for lossless MTL, 210, 212
SPICELPI.FOR,SPICELPI.EXE
  computer programs, 227
SPICEMTL.FOR,SPICEMTL.EXE
  computer programs, 216
SPICEMTL.IN, 219
square wave, 14
STRPLINE.FOR,STRPLINE.EXE, 204
superposition in linear systems, 27

the .AC execution statement
  in SPICE, 282
the .DC execution statement
  in SPICE, 282
the .FOUR Fourier statement
  in SPICE, 284
the .PRINT output statement
  in SPICE, 283

the .TRAN execution statement
in SPICE, 282
the basic transmission-line problem, 32
the periodic pulse train (PULSE) source
waveform
in SPICE, 281
the piecewise linear (PWL) source waveform
in SPICE, 281
the sinusoid source waveform (SIN)
in SPICE, 281
the subcircuit model (SUBCKT) statement
in SPICE, 292
three-conductor lines
PCB lands, 180
wires, 179
three-conductor lossless lines
mode transformations, 208
per-unit-length parameter
matrices, 208
uncoupled transmission-line equations,
207
three-conductor lossy lines,
MTL transmission-line equations, 242
time delay, 7
time domain vs frequency domain, 11
time-domain of two-conductor lines, 51
time-domain to frequency-domain method,
261
TIMEFREQ.FOR,TIMEFREQ.EXE, 266
transmission-line equations
general solution, 56
two-conductor lines, 53
uncoupled, 55
three-conductor lines, 177
transmission lines (lossless)
in SPICE, 280

transverse electromagnetic (TEM) mode of
propagation, 52
Transverse ElectroMagnetic (TEM)
waves, 11
two-conductor lossy line transmission-line
equations, 241
uncoupled phasor MTL equations, 244
uncoupled phasor transmission-line
equations
two-conductor lines, 122

uniform line, 52
unit multipliers, 4
units and unit conversion, 3

velocity of propagation, 7
voltage and current along the line, 130
voltage reflection coefficient, 71
phasor solution, 123

wavelength, 7
waves
backward-traveling, 6
form of, 6
forward-traveling, 6
wide-separation approximations for
wire-type MTLs
two wires within an overall
shield, 190
WIDESEP.FOR, WIDESEP.EXE
computer programs, 192
wide-separation approximations for
wire-type MTLs, 185
three wires, 187
two wires above a ground plane, 188
two wires in an overall shield, 190

Printed and bound by CPI Group (UK) Ltd, Croydon, CR0 4YY

07/12/2023